T0226234

Mechanik, Werkstoffe und Konstruktion im Bauwesen

Band 60

Reihe herausgegeben von
Ulrich Knaack, Darmstadt, Deutschland
Jens Schneider, Darmstadt, Deutschland
Johann-Dietrich Wörner, Darmstadt, Deutschland
Stefan Kolling, Gießen, Deutschland

Institutsreihe zu Fortschritten bei Mechanik, Werkstoffen, Konstruktionen, Gebäudehüllen und Tragwerken. Das Institut für Statik und Konstruktion der TU Darmstadt sowie das Institut für Mechanik und Materialforschung der TH Mittelhessen in Gießen bündeln die Forschungs- und Lehraktivitäten in den Bereichen Mechanik, Werkstoffe im Bauwesen, Statik und Dynamik, Glasbau und Fassadentechnik, um einheitliche Grundlagen für werkstoffgerechtes Entwerfen und Konstruieren zu erreichen. Die Institute sind national und international sehr gut vernetzt und kooperieren bei grundlegenden theoretischen Arbeiten und angewandten Forschungsprojekten mit Partnern aus Wissenschaft, Industrie und Verwaltung. Die Forschungsaktivitäten finden sich im gesamten Ingenieurbereich wieder. Sie umfassen die Modellierung von Tragstrukturen zur Erfassung des statischen und dynamischen Verhaltens, die mechanische Modellierung und Computersimulation des Deformations-, Schädigungs- und Versagensverhaltens von Werkstoffen, Bauteilen und Tragstrukturen, die Entwicklung neuer Materialien, Produktionsverfahren und Gebäudetechnologien sowie deren Anwendung im Bauwesen unter Berücksichtigung sicherheitstheoretischer Überlegungen und der Energieeffizienz, konstruktive Aspekte des Umweltschutzes sowie numerische Simulationen von komplexen Stoßvorgängen und Kontaktproblemen in Statik und Dynamik.

Weitere Bände in der Reihe https://www.springer.com/series/13824

Evgenia Kanli

Experimental Investigations on Joining Techniques for Paper Structures

A Showcase of Principles, Case Studies & Novel Connections Created in the Spirit of Architectural Engineering

Evgenia Kanli
TU Darmstadt
Darmstadt, Germany

Vom Fachbereich 13 - Bauingenieurwesen und Umweltwissenschaften der Technischen Universität Darmstadt zur Erlangung des akademischen Grades eines Doktor-Ingenieurs (Dr.-Ing.) genehmigte Dissertation von Evgenia Kanli, M.Sc. aus Thessaloniki, GR

1. Gutachten: Prof. Dr.-Ing. Ulrich Knaack
2. Gutachten: Prof. Dr.-Ing. Tilmann Klein
Tag der Einreichung: 5. Januar 2021
Tag der mündlichen Prüfung: 2. März 2021

Darmstadt 2021

ISSN 2512-3238   ISSN 2512-3246   (electronic)
Mechanik, Werkstoffe und Konstruktion im Bauwesen
ISBN 978-3-658-34500-6   ISBN 978-3-658-34501-3   (eBook)
https://doi.org/10.1007/978-3-658-34501-3

This Springer Vieweg imprint is published by the registered company Springer Fachmedien Wiesbaden GmbH part of Springer Nature.
The registered company address is: Abraham-Lincoln-Str. 46, 65189 Wiesbaden, Germany

Abstract

The background of this research is related to innovative lightweight construction methods for short-term applications realized with highly recyclable materials produced from renewable resources. In this spirit, the potential utilization of paper-based products for construction purposes is examined. Previous experiences indicate that even though a variety of concepts for designs is offered, mostly as a result of research in architecture and relevant workshops, there is only a limited number of constructed projects. The lack of expertise in detailing, planning and building full-scale prototypes has been a significant drawback. To this respect, the integral subject of connections remains up to this point vague, meaning the state of the art, the functionality and the actual limits of consolidated solutions. This research focuses on the subject of joining techniques, aiming to shed some light on these aspects.

The research questions regard the state of the art and future potential of joining techniques for paper-based products and structures (1), the distinct preference on paper-tubes for structural purposes and the joining techniques that apply in this case (2) and the identification of effective joining methods for this application based on certain criteria and priorities (3).

(1) The joining principles are clustered in two categories, basic techniques and specialized solutions for modern paper-products such as lightweight boards and profiles. In continuation, crosslinks with assemblies from common construction materials (timber, bamboo, textiles, composites and steel) are identified with the aim to point out influences towards paper-based structures. The result of this process is a global overview of possibilities for joining paper-based products in an attempt to visualize the future potential in the field of construction.

(2) In the second section of the research the prevalent construction typologies are briefly presented and the focus is set on skeletal structures that use paper-tubes as the main structural elements. The dominance of this construction method in the field of paper-structures is distinct within section one (1). The connection points between the primary structural members are the main subject set for the elaboration of further research. The distribution of loads with global FEM analysis, in 'RFEM Dlubal' software, is examined for three typical cases of beam structures, to observe the issues and the boundary conditions for the functionality of the joints. Next to this, the features of the paper-tubes selected for the research process are clarified.

(3) In continuation, a showcase of design concepts is presented and a selection of concepts for further development is pursued. These include in principle form-locking, plug-in and sleeve connectors. Use of steel elements and adhesives apply according to the

special needs of the individual studies. For each case study the following areas are discussed: characteristic joining method, detailing alternatives, assembly methods, manufacturing possibilities and structural performance. To investigate the manufacturing process several series of prototypes are built. To study the structural performance targeted structural tests are performed, primarily bending tests. In this context the opportunity to examine basic joining principles as part of the hybrid joints arises (such as screws and bolts, adhesives), as well as to track and quantify the performance of the joints. The results from the structural tests are used to discuss representative values, such as the bending strength and stiffness of the specimens and when possible the rotational stiffness of the joint. Next to this, local FEM analysis, in 'ANSYS' software, is performed with the main target to analyze the distribution of stresses in the different assemblies. The main advantage of the local FEM analysis comparing to the structural bending tests is the possibility to implement a combination of loads on a multi-axial joint. For this purpose, input from the global FEM analysis is used, in order to simulate realistic boundary conditions. In addition to this, comparisons between the structural testing and the FEM analysis are made, focusing on the critical areas and the failure mechanisms. The case studies are evaluated on all aforementioned levels and comparisons between the different joining techniques are drawn.

Overall, this thesis starts by informing the current status on the subject of joining techniques for paper-based products and structures and continuous with organized efforts in examining comprehensively selected cases of multi-axial joints as parts of skeletal structures. Based on the research process, the structural issues seem to be particularly critical for the feasibility of the joints. In this area the findings indicate both the advantages of plug-in wooden joints for strength and the superiority of sleeve joints based on adhesive bonding for stiffness despite, in this case, the deficits for strength and warnings prior to structural failure. The category of sleeve joints presents great potential on many levels but further development of the manufacturing processes is required to achieve the desired results.

A persistent dilemma that is present in various stages of the research development, when considering constructing with paper-based materials, is related to the preservation of the 'cradle to cradle' strategy. To explain this a bit further, in this particular research, it often comes down to the possibility of reinforcing the construction elements by using strong highly absorbent adhesives, to achieve satisfactory increase in strength and durability of the assembly. In the end, the motivation of this research is to examine an alternative for temporary structures with sustainable character and therefore this issue cannot be overlooked.

Zusammenfassung

Der Hintergrund dieser Forschungsarbeit bezieht sich auf innovative Leichtbauweisen für vorübergehende Konstruktionen, die aus hoch recyclebaren Materialien aus nachwachsenden Rohstoffen gebaut werden. In diesem Zusammenhang wird die potentielle Verwendung papierbasierter Produkte zu Konstruktionszwecken untersucht. Bisherigen Erfahrungen zufolge steht zwar eine Vielzahl von Baukonzepten, meist als Ergebnis von Forschungen in der Architektur sowie einschlägiger Workshops, zur Verfügung, jedoch wird nur eine begrenzte Anzahl jener Konzepte umgesetzt. Die fehlende Expertise bei der Detaillierung, der Planung und dem Bau maßstabsgetreuer Prototypen ist ein wesentliches Manko. Insbesondere bleibt bis heute das grundlegende Thema der Verbindungsmöglichkeiten, d.h. der Stand der Technik, die Funktionalität und die tatsächlichen Grenzen konsolidierter Lösungen, vage. Die vorliegende Arbeit befasst sich mit dem Thema Fügetechniken und versucht, diese Aspekte zu beleuchten.

Die Forschungsfragen beziehen sich auf den Stand der Technik und das zukünftige Potential von Fügetechniken für papierbasierte Produkte und Strukturen (1), die eindeutigen Vorzüge von Papierrohren für Konstruktionszwecke und die in diesem Fall anzuwendenden Fügetechniken (2) und die Identifizierung effektiver Fügeverfahren für diese Anwendung anhand bestimmter Kriterien und Prioritäten (3).

(1) Die Fügeprinzipien werden in zwei Kategorien, Basistechniken und spezielle Lösungen für moderne Papierprodukte wie u.a. Leichtbauplatten und Profile, eingeteilt. Ferner werden Querverbindungen zu Bauteilen aus konventionell Baumaterialien (Holz, Bambus, Textilien, Verbundwerkstoffe und Stahl) mit dem Ziel identifiziert, Einflüsse auf papierbasierte Strukturen aufzuzeigen. Das Ergebnis dieses Prozesses ist ein globaler Überblick über die Möglichkeiten des Verbindens papierbasierter Produkte, um das zukünftige Potential im Bereich des Bauens zu visualisieren.

(2) Im zweiten Abschnitt der Arbeit werden die gängigen Bautypologien kurz vorgestellt, wobei der Schwerpunkt auf Skelettkonstruktionen gelegt wird, welche Papierrohre als Hauptstrukturelemente aufweisen. Die Dominanz dieser Bauweise im Bereich der Papierkonstruktionen ist im ersten Abschnitt (1) deutlich zu erkennen. Die Verbindungspunkte zwischen den Hauptstrukturelementen sind das zentrale Thema für die Ausarbeitung der weiteren Forschung. Die Lastverteilung mit globaler FEM-Analyse mithilfe der Software 'RFEM Dlubal' wird für drei typische Fälle von Balkenkonstruktionen untersucht, um die Problematik und die Randbedingungen für die Funktionalität der Verbindungen zu beobachten. Daneben werden die Eigenschaften der für den Forschungsprozess ausgewählten Papierrohre erläutert.

(3) Im weiteren Verlauf werden Designkonzepte vorgestellt und eine Auswahl von Konzepten für die weitere Entwicklung untersucht. Dazu gehören im Prinzip formschlüssige, steckbare und hülsenförmige Verbindungsstücke. Die Verwendung von Stahlelementen und Klebstoffen erfolgt entsprechend den speziellen Anforderungen der jeweiligen Studien. Daneben werden pro Fallstudie folgende Bereiche diskutiert: charakteristische Fügeverfahren, Detaillierungsalternativen, Montagemethoden, Fertigungsmöglichkeiten und Tragwerksperformance. Zur Untersuchung des Herstellungsverfahrens werden mehrere Serien von Prototypen gebaut. Zur Untersuchung der strukturellen Leistungsfähigkeit werden gezielte Strukturtests, vor allem Biegeversuche, durchgeführt. In diesem Zusammenhang ergibt sich sowohl die Möglichkeit, grundlegende Verbindungsprinzipien als Teil von Hybridverbindungen zu untersuchen (z. B. Schrauben und Bolzen, Klebstoffe), als auch die Leistungsfähigkeit der Verbindungen zu verfolgen und zu quantifizieren. Anschließend werden anhand der Ergebnisse der Strukturversuche repräsentative Werte diskutiert, wie z.B. die Biegefestigkeit und Steifigkeit der Probekörper und wenn möglich die Rotationssteifigkeit der Verbindung. Daneben wird eine lokale FEM-Analyse in der Software 'ANSYS' mit dem Hauptziel durchgeführt, die Spannungsverteilung in den verschiedenen Baugruppen zu analysieren. Der zentrale Vorteil der lokalen FEM-Analyse im Vergleich zu den strukturellen Biegeversuchen ist die Möglichkeit, eine Kombination von Belastungen auf ein mehrachsiges Gelenk zu implementieren. Dazu wird der Input aus der globalen FEM-Analyse verwendet, um realistische Randbedingungen zu simulieren. Darüber hinaus werden Vergleiche zwischen den Strukturversuchen und der FEM-Analyse durchgeführt, wobei der Fokus auf den kritischen Bereichen und den Versagensmechanismen liegt. Die Fallstudien werden in allen vorgenannten Ebenen ausgewertet und Vergleiche zwischen den verschiedenen Fügetechniken gezogen.

Insgesamt informiert diese Arbeit zunächst über den aktuellen Stand zum Thema Fügetechniken für papierbasierte Produkte und Strukturen und setzt sich dann in organisiertem Bestreben fort, ausgewählte Fälle mehrachsiger Verbindungen als Teil von Skelettstrukturen umfassend zu untersuchen. Basierend auf dem Forschungsprozess erscheint das Thema Struktur besonders bedenklich für die Umsetzbarkeit der Knoten zu sein. In diesem Bereich zeigen die Ergebnisse sowohl die Vorteile von steckbaren Holzverbindungen für die Festigkeit als auch die Überlegenheit von Manschettenverbindungen auf Basis von Klebeverbindungen für die Steifigkeit, trotz der im letzten Fall vorhandenen Defizite im Hinblick auf Festigkeit und Warnhinweisen vor strukturellem Versagen. Die Kategorie der Manschetten-verbindungen birgt in vielerlei Hinsicht ein großes Potenzial, jedoch ist eine Weiterentwicklung der Herstellungsverfahren erforderlich, um die gewünschten Ergebnisse zu erzielen.

Ein anhaltendes Dilemma, das in verschiedenen Stadien der Forschungsentwicklung auftritt, wenn es darum geht, mit papierbasierten Materialien zu konstruieren, steht im Zusammenhang mit der Einhaltung der ‚Cradle to Cradle'-Strategie. Um dies etwas näher zu erläutern, geht es in dieser speziellen Forschung oft um die Möglichkeit, die Konstruktionselemente durch die Verwendung starker, hoch absorbierender Klebstoffe zu verstärken, um eine zufriedenstellende Erhöhung der Festigkeit und Haltbarkeit der Baugruppe zu erreichen. Letztendlich ist die Motivation dieser Arbeit, eine Alternative für temporäre Strukturen mit nachhaltigem Charakter zu untersuchen und daher kann dieses Thema nicht übersehen werden.

Acknowledgments

Throughout the process of developing this research some people, both from my personal and professional life, helped me to make progress and at times overcome difficulties. I would like to take a moment to thank them for their contribution and express my gratitude for being lucky enough to have met them.

I have been blessed with a double alter ego, my mother Margarita, a passionate teacher, who tirelessly embraced all my efforts to learn and develop and grandmother Eugenia, an incredible craftsman, who taught me how to be creative in the making. Their inspiration is always a great source of energy and a compass for me. I am grateful to Robert for being a solid support and encouraging diverse perspective throughout the whole research process. I am grateful to my colleague Nihat for encouraging my efforts in research, sharing his knowledge on structural mechanics and material testing and more than that looking out for me.

I would also like to thank my colleagues and co-researchers who helped me to approach areas that were new to me. Specifically, I would like to thank Paul for getting on board with my attempts to try-out a special connection and offering his expertise (see p. 261). From a different perspective, I am thankful for the students who were curious about my research, gave me the opportunity to cultivate my teaching skills and discover how fulfilling this process is for me. I would like to thank Noah (see p. 274), for travelling all the way from Colorado (US) and diving in the research with such a positive attitude, providing his perspective as a student from mechanical engineering.

I would like to thank my mentors for supporting this process and the Institute of Structural Mechanics and Design from TU Darmstadt for giving me the space and opportunity to conduct my research leading to this dissertation.

This research acknowledges financial support within the LOEWE program of excellence of the Federal State of Hessen (project initiative BAMP!).

Contents

Glossaries

Abbreviations

Case studies

The following abbreviations, for the names of the case studies elaborated in this research, were used during the processes of prototyping and performing structural experiments.

CW   Crossed-wood, for ‚Timber plate puzzle node‘

MW   Massive wood, for ‚Timber block node‘

TAJ   Tolerance adaptive joint

BBC   Bio-based clamp, for 'Fiber reinforced pulp-composite connector'

LAMT   Laminated textile, for 'Textile-reinforced epoxy resin laminated joint'

General

CAD   Computer aided design

CNC   Computer numerical control

FEM   Finite Element Method

LC   Load combination

Materials

CLT   Cross laminated timber

UHPC   Ultra-high-performance concrete

Symbols

Global structural analysis

d	Displacement (mm)
E_m	Modulus of elasticity (kN/m^2)
F	Force (N)
f	Critical factor (output from RSBUCK module in RFEM software)
$U_{t,max}$	Maximum total deformation (mm)
γ	Specific weight (kN/m^3)
γ_M	Partial safety factor
ν	Poisson's Ratio
σ_{yield}	Stress at yield point (MPa)

Structural testing & interpretation of results

B	Bending stiffness of specimen (kN/mm)
dl_A	Displacement at Point A (support - as measured by 'Zwick/ Roell')
dl_B	Displacement at Point B (as measured by displacement-sensor)
$F_{max,\ Linear}$	Maximum force (limit before plastic deformation) (kN)
$F_{max,\ Ultimate}$	Maximum force (ultimate limit, before failure) (kN)
M_{max}	Maximum moment of specimen (kNmm)
σ_{max}	Maximum stress of specimen (MPa)
$\sigma_{\varphi y}$	Spring constant of specimen (Nm/°)

Introduction

Modern industrial production in the field of paper products has advanced significantly and has delivered components with qualities that present potential for new applications. In the past decades there have been attempts to use paper-based materials for construction purposes. These experiences provide some interesting examples that indicate the potential for researchers to explore a collection of issues. Therefore, it is essential to bring paper-based materials in the same picture with construction materials and examine the possible outcomes of this venture.

The subject of joining techniques is of great importance, in order to bring designs and structures to life and develop effective assemblies. Especially due to the experimental character of this idea, it is found appropriate to identify ways to approach this possibility and examine its feasibility. This research aims to create groundwork on this subject, from the perspective of architecture and building technology.

The research is performed in the following three main phases: (1) The review of known joining techniques for paper-based products (state of the art) and the identification of further principles for joining techniques that present potential for implementation on paper-based assemblies, (2) a comprehensive analysis of the construction outlines that are selected as a target for the development of case studies, (3) the presentation of the process and findings produced in the targeted case studies that are executed, with regard to the subject of multi-axial nodes for paper-based round profiles, considering particularly the aspects of detail, assembly, manufacturing and structural integrity for the joints.

In phase 1 (chapter 2) the entire spectrum of joining techniques for the total of paper-based products and structures is analyzed. Significant efforts are made to categorize and visualize the most relevant techniques and communicate effectively the current possibilities. After establishing the state of the art, scenarios for further development are conceived by identifying crosslinks with established construction materials. The whole process is embodied in a global overview that concludes this phase.

The second phase (chapter 3) aims to create a smooth transition between the global overview of joining techniques (chapter 2) and the case studies examined (chapter 4). This way, in chapter 3, the prevalent typologies of paper-based structures are briefly discussed and typical cases of beam-structures composed of paper-based round profiles are examined, with respect to the principle construction outlines and global structural analysis. Hence, the main subjects explored are these of potential structural applications and, to a further extent, the boundary conditions for the joining techniques that are investigated in the next phase of the research.

In the third phase (chapter 4), the focus is drawn exclusively on the product of paper-tubes that up until now is the most highly preferred paper-based product, for the purpose of building load bearing structures, based on the findings from phase (1), as explained in paragraphs 2.1.1 'Collection of reference projects and studies' and 2.4 'Epilogue for the state of the art and direction of the research following'. This phase is devoted to experimental research work that includes the use of modern tools, such as software for computer aided design, a variety of fabrication methods for the generation of prototypes, structural tests and computational simulations of the structural performance. A number of case studies for multi-axial joints is developed and evaluated mainly on the levels of design flexibility, manufacturing process, effective assembly, structural performance and durability. This process is summarized by the end of chapter 4, where comparisons between the case studies are drawn.

The experimental research process of building prototypes, performing structural tests, as well as attempting simplified simulations in FEM analysis, was intended to create new experiences in engineering solutions and to generate, as far as possible, consolidated conclusions.

The findings of the whole research are compiled in chapter 5, according to the research questions as these are set in chapter 1.

Figure 1.1 Microscopic view of paperboard (ISM+D)

1 Research methodology

Keywords: research methodology, criteria

Hereby the research outlines are set. The motivation and main focus of the research are explained. The research questions and main objectives are stated. The main parts of the process and milestones are defined. In continuation, the research methods implemented are analyzed and further considerations are discussed.

1.1 Motivation and main focus

The use of paper-based products in construction is a possibility under investigation, still a novel topic and a relatively unknown field both for engineers and designers. In this respect, it's mainly a subject of research that up until now has mostly found implementation in temporary structures, such as shelters for humanitarian purposes and temporary installations for events. There are very few examples which regard more long-term applications.

Numerous challenges need to be overcome, to explore this new experimental building technology, as there is not enough information yet about the functionality of structures of this kind. Thus, it is important to identify and shape further the context for the potential implementation of paper products in design and construction products, to inform the targeted applications and eventually study the feasibility of the researched cases.

The aspect of joining techniques, as a combination of design and engineering oriented perspective, is the main focus of this research. Below, the most important issues that shape the motivational background are described.

- Outlining the global potential

Existing studies in the broader field of architecture and civil engineering highlight the value of detailing for the joining techniques, for different assemblies and functions. In some cases, the aspect of joining techniques is briefly discussed, creating some precedents for this research subject as it has evolved in the last two decades.

At the same time, up to this point, various approaches for the subject of joining techniques for paper-based structures can be observed, based on the priorities set by the primary subject of research. This way, the subject of joining techniques for paper-based structures has not been yet in the main focus, on its own. It mostly remains as a follow-

© The Author(s), under exclusive license to Springer Fachmedien Wiesbaden GmbH, part of Springer Nature 2022
E. Kanli, *Experimental Investigations on Joining Techniques for Paper Structures*, Mechanik, Werkstoffe und Konstruktion im Bauwesen 60, https://doi.org/10.1007/978-3-658-34501-3_1

up topic for investigation. Specific examples of groundwork and reference projects are reported in the state of the art (chapter 2). Through the different researches, the questions created on aspects related to the joining techniques are numerous, yet fundamental issues regarding the applicability of referred solutions remain vague. Furthermore, in many cases of published researches, the legal issues for the realization of paper-based structures are addressed and the necessity for conduction of thorough investigations to examine the functionality of assemblies and to test connections between structural components is highlighted.

In this greater context, this research aims to take the subject of joining techniques as the starting point. Fundamentally, the extensive discussion of potential joining techniques for different groups of paper-based products is the first priority, as well as predictable deficits. The global aim is to present arguments that indicate the effectiveness of potential joining principles. The in-depth investigation of selected case studies, to prove the functionality of joining techniques under specific conditions, follows.

In this attempt, it is considered important to develop a strategy in mapping out joining principles, in a way which corresponds with the fundamental joining principles that apply for conventional materials. The higher motivation is to improve the way the global research subject is communicated within the global field of construction.

- Paper-tubes and multi-axial joins

As it is explained extensively in the start of chapter 2, based on existing projects, high preference in specific construction typologies is expressed, when building with paper-based components. At the moment and for the past decades the product of paper tubes has been the most popular paper-based component with application in construction, as explained in paragraph 2.1. To briefly argue this observation, based on the review of realized constructions and especially the ones built outdoors, the structures that include paper tubes in the group of load-bearing components, versus composite elements composed of paper-based boards, are predominant. Even though it is hard to make an exact estimation about the number of realized paper-structures, due to the fact that a number of experimental prototypes are not published, there are valid reasons for this preference. A fundamental reason lies with the simplicity in adjusting basic characteristics of the paper-tubes within the production process, such as the base material, the diameter and wall thickness, so that the finished components can be directly used, without further processing. Next to this, the round shaped profiles present structural advantages and in combination with the solid wall, the structural capacity particularly against axial compression reaches promising performances. Specific information about the structural performance of the paper-tubes is provided in paragraph 3.1.2. On the other hand, paper-based solid or hollowed boards (with corrugated or honeycomb core, used as an intermediate layer in the sandwich structure), require further processing, for the purpose of manufacturing components with sufficient geometrical characteristics and structural integrity. As there are no such standard components, this concept has been, so far, less elaborated, especially in professional environments.

Regarding the subject of structures composed of paper-tubes, despite the existence of realized projects, the majority of the technical details and engineering data remain unpublished. To a different extent, these projects are realized in different places, times and by different groups. Hence, systematic research and evaluation of solutions that would clarify the functionality of specific joining techniques are currently missing.

- Prevalent structural issues following the material composition

Paper-based products are composed of cellulose fibers with length somewhere between 2 and 3.5 mm (iggesund, 2020). The principal inhomogeneous material structure is a primary issue and often the main cause of structural failure of paper-based components. The integration of adhesives in composite multi-layered components challenges the design of the joints further. Local stresses intensify this problem, due to the distribution of higher amounts of force in a small area that puts the bond-strength, between the individual layers and fibers, to test. In total, the unpredictability of the structural performance of the components, in combination with the softness and brittleness of the material itself pose doubts. This way, it is important to make steps towards obtaining information about how close functional assemblies are to satisfactory performances.

- The design challenge

The design and materialization of the joints are crucial aspects for the functionality of the construction. Considering that in the areas of the joints, there is typically higher concentration of stresses, the structural issues are highly intensified. The possibility of using the design of the joints to improve the functionality of the assembly is interesting. For example, as commented in a review about the use of paper-tubes in structural engineering (Bank, 2016) and particularly truss-like structures, the design of the nodes is perhaps the most crucial aspect for the success of the whole construction. This way, a rich subject within these investigations lies on studying different design outlines for the same principle assembly, examining their functionality and drawing comparisons.

- A multidisciplinary research initiative

In the last two decades, the potential implementation of paper-based products in building applications has been examined by numerous researchers. These efforts have generated motivation for the present research, to investigate further aspects that are important for 'paper architecture'. This is also one of the global aims of the research project 'Building with paper' (BAMP!, 2017). In this respect, the research on joining techniques is an integral part of the aims that regard construction related issues, with a further view on the design possibilities and potential applications. A summary of the aspects discussed in this dissertation is provided to be presented in a book-edition of an 'Atlas' that aims to combine discussions from researchers involved in this project.

1.2 Research questions

The research questions are focused entirely on the subject of joining techniques, for assemblies composed of paper-based products. The attention is mainly drawn to the subjects of (1) the global state of the art, including the analysis of further potential and (2) a decisive approach on targeting structurers built with paper-based tubes, as a subject of greater focus and (3) a suitable context for performing experiments.

The research questions are stated below. The three main research questions aim to identify the principle direction of this research, whereas the sub-questions to add further details for the scope and deliverables of the research.

(1) What is the 'state of the art' of joining principles for experimental structures built primarily with paper-based materials?

    a. Which are the main categories of joining principles that can be identified within the 'State of the art' and what are the main pros and cons for implementation in construction?

    b. How could the 'State of the art' of joining principles develop in the future, considering further potential solutions?

For the question 1b it is important to mention the considerations about potential links between common assembly methods and assemblies performed with paper-based products. In other words, the discussion of pinpointing common ground and suggesting compatible and incompatible qualities between paper-based products and common building materials.

(2) Considering that the product of paper tubes is the most highly applicable paper-based component in construction so far, particularly for load-bearing elements, what are the most prevalent joining techniques implemented and why?

    a. As part of exploring the potential implementation of paper-based products in construction, how can the preference to paper-tubes comparing to other products be explained and what other types of structures that make use of other paper products can be identified?

    b. Exploring the potential of this construction typology, what structural configurations and applications are suitable, based on the characteristics of the paper-tube? Next to this, which joining methods are suitable (also in combination with research question 1)?

(3) Which joining techniques show the best potential for small scale tubular structures, to be used as a temporary living space, considering the aspects of design, assembly, stability, materialization and production?

a. What possibilities shall be further explored, as case studies, in accordance with the findings following research questions 1 and 2?

b. What are the greatest difficulties in designing suitable joints for tubular structures, especially considering the requirements for structural integrity? Continuously, what are the most critical aspects for the structural integrity of the joints for the different cases examined?

c. Classify the pros and cons identified for the case studies elaborated within this research, considering the main aspects addressed in question 3 and also other aspects that might appear to be important as indicated by the research process.

1.3 Objectives

The global objective is to examine the subject of joining techniques for paper-based structures as a separate issue within the corresponding field of research. Following the research questions a set of objectives is derived that requires a combination of different efforts and research methods to achieve the aims appointed for the research. These regard mainly the 'state of the art' about joining techniques for paper-based assemblies, the structural applications of paper-tubes and the joining techniques that are suitable in this particular category of paper-based structures. In this process it is important to demonstrate the relations between the joining techniques considered and the corresponding paper-products. Next to this, it is important to identify how to use the research tools and methods available to create further insights for this purpose. Below the main objectives as a result of the research questions are clarified:

(1) Following research questions (1) and (2): Generate an overview of joining techniques that are justified for paper-based components, assemblies and structures.
   a. Review existing solutions
   b. Identify solutions from common construction that present potential for paper-based structures
(2) Following research question (2b):
   a. Review representative structural concepts
   b. Examine closely a selection of structural concepts and set the context for the experimental case studies
(3) Following research question (3a):
   a. Create design concepts and make a systematic selection
   b. Develop case studies and frame a comparative method

(4) Following research question (3b), in combination with the global principle requirements:

    a. Examine the key-aspects of design, detailing alternatives, production techniques and structural performance per case study in order to identify the pros and cons.

    b. Create proof for the functionality of the joints in relation with the aspects stated in (4a).

The further aim of this effort is to inspire designers and engineers, who might be interested in such innovative research topics, by presenting an experimental approach that attempts to merge creativity with the methodical research process.

1.4 Process and milestones

Table 1.1 The main process, milestones and global objectives of this research.

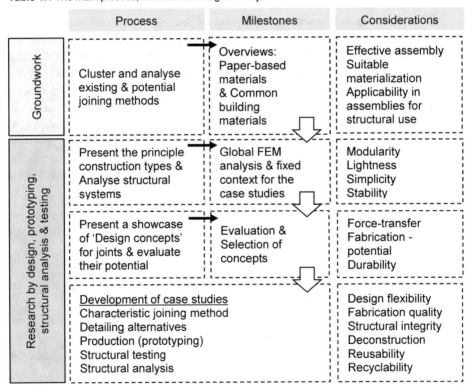

The research is developed on different levels and scales, to achieve the desired results.

In this spirit, the research process is primarily divided in two stages. The first stage combines theoretical instruments and precedents to depict the state of the art and its potential. This stage considers a wide variety of paper-based products and their potential integration in assemblies. The second stage aims to expand the current knowledge, by performing new research and experiments with a special target. In this stage, the focus lies on the product of paper-based tubes, following the motivation of the research.

This way, the research process is mainly divided in the two following sections:

(1) The analysis of the 'state of the art' and identification of potential solutions. The approach of clustering joining techniques for paper-based products is a focal point. Realized paper-structures are included in this process, to showcase a variety of joints that have already been implemented for construction purposes.

(2) The development of experimental case studies on multi-axial nodes for beam-structures composed of paper-tubes.

1.5 Research tools

Hereby, the main research tools and methods used to evaluate the findings are introduced. These mainly regard: the process of reviewing existing joining techniques and the strategy set for outlining further potential, the experimental approach followed within the elaboration of case studies and the approach for the interpretation of the findings. The accessible facilities at the 'Institute of structural mechanics and design', influenced the direction of the research. The available workshop for the development of prototypes, the equipment for performance of structural tests and the software for structural simulations were viewed as opportunities to shape an interesting research approach that can lead to new experiences. Ultimately, the perspectives of design and engineering are both incorporated and interweaved in the research process.

1.5.1 Analysis and review of existing joining techniques

This analysis takes place in chapter 2.1. As a starting point, a brief review of the most relevant paper-based products is provided. The focus is on addressing the forming process and the possible geometrical characteristics of the finished components. For example, looking at the paper products drawn in fig. 1.2 it is easy to understand that they are all produced with different processes. This fact has great influence on the structural behaviour of the finished components and consequently on the design of assemblies and joints.

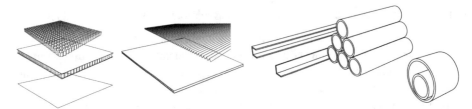

Figure 1.2 Paper-based products, boards and profiles, that present interest for design applications. Namely from left to right: honeycomb board, corrugated paper, various paper-board profiles (round, L and U).

A variety of joining techniques is being reviewed and analyzed regarding the design, assembly method and functionality. For this purpose, the review is divided in sub-sections as shown in fig. 1.4. Overall, within each sub-section the joining techniques are mainly clustered in the following four groups: material closure, force closure, form-locking and hybrid solutions. This distinction is inspired by (VDI, 2004).

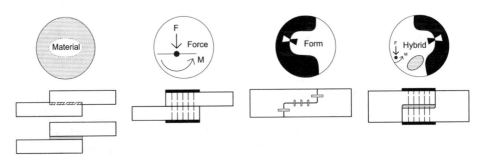

Figure 1.3 The four main clusters of joining techniques: material-closure, force-closure, form-locking and hybrid solutions.

The criteria for the selection of the references that are included in the review, are explained in detail in paragraph 2.1. For this research, in principle, the focus lies on professional construction work and published research. To mention a few representative examples, temporary construction projects are developed by the offices of the architect Shigeru Ban (JPN), who is known for the 'humanitarian architecture projects', as well as Octatube (NL) and ABT consulting engineers (NL) that have developed interesting construction details using mainly paper-tubes as structural members. In the field of research, as mentioned in the motivation (1.1) some European Institutes that have developed research on the subject of using paper materials for construction purposes, in recent times, are ETH Zurich (Pohl, 2009), TU Delft (Latka 2017) and Bauhaus-Universität Weimar (Schütz, 2017). In total, there are numerous projects occupied with paper design and architecture, especially in the fields of interior design and emergency shelters and in creative labs or educational environments. For the majority of these projects, the building systems, assembly methods and joining techniques presented are often rather similar.

To outline further potential solutions, a parallel review of selected joining techniques from materials that are commonly used for building purposes takes place. The materials selected for this reason are the following: timber – due to the shared material origins –, textiles – as they are commonly thin, soft and present advantages for tensile forces, like a paper-sheet –, composites – by reason of certain processing methods for the integration of fasteners in multi-layered composites –, bamboo – because as a component used in structural frames, it is interesting to observe parallel to the paper-tubes, – and metal – as joining techniques for this material are perhaps the most elaborate solutions in construction in total. The literature sources for joints in common construction include primarily construction manuals, and standards (for example Eurocode 5 for timber structures (CEN, 2004), DIN8593 etc.).

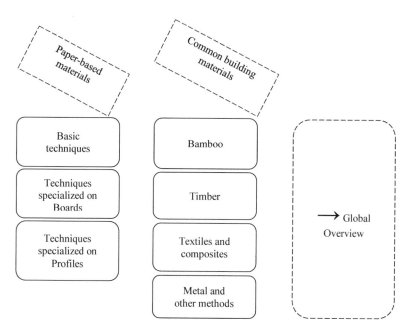

Figure 1.4 Concept for the review of existing solutions

By the end of chapter (2) an overview of joining techniques is provided. This is accompanied by informative tables that indicate the pros and cons per technique. The following criteria are encountered as indicators for the functionality of principle joining techniques: transferability of forces, technological means required for the fabrication process, critical aspects that could lead to failure and recommended applications. Additionally, the aspects of deconstruction and re-usability are discussed within the accompanying conclusions. The selection of the criteria was made following observations on the structural issues of paper-based materials with suspected impact on the areas of the joints and the priorities about the aspects of practicality and feasibility of the suggested assemblies.

1.5.2 Global structural FEM analysis

In chapter 3.2 three typical cases of skeletal structures are selected to demonstrate possible contexts for the experimental investigations on joining techniques (case studies, chapter 4). These are an orthogonal grid, a dome and an arch-tunnel. The aim is to analyze the distribution of loads in each structure, identify the most critical conditions and also draw a line for the required structural performance of the joints (benchmark).

The structural behavior of the structures is examined by performing global FEM analysis. The software used for this purpose is 'Dlubal RFEM'. It provides the possibility to study the distribution of loads on 3-dimensional objects. The structural model is composed of members and nodes. The software provides an extensive library of materials and conditions for the application of loads as well as calculation methods. Moreover, it offers 'stability analysis of beam structures', with the additional module named 'RSBUCK' that indicates how critical buckling is for the input provided.

A significant challenge in using this software is that the input required for the mechanical properties of the beams is customized and simplified. As expected, paper-based materials are not included in the material library and the standards incorporated in the software. Hence, this method is used to roughly approximate the global structural behavior. The input regarding all structural members and load cases is extensively explained in chapter 3.2.

1.5.3 Case studies: aspects and research tools

In paragraph (4.1) a collection of design concepts is considered, from which a few are selected for further development. The main selection criteria adhere to the criteria applied in the end of chapter (2) – overview of joining principles – and include the following aspects: transferability of forces, manufacturing process (automated/ handcrafted), relative cost (raw materials/ production), assembly/ disassembly process, durability and recyclability. Further details are provided in paragraph 4.1.1.

The next methods regard the development of the individual case studies.

(1) Analysis of characteristic joining method

A method to systematically describe the matrix of a joint is presented in (VDI, 2004). For each case study, the main relations between the assembly members are visualized, in the spirit of the aforementioned reference and also following the principles shown in fig. 1.3.

(2) Investigation of detailing alternatives

Different scenarios are sketched. This method is used to evaluated the flexibility of the design when needed to be adjusted, either for different structural or geometrical requirements.

(3) Production techniques

In principle, the investigation of production techniques partly refers to a trial and error approach with multiple iteration steps to achieve optimized results. The three main aspects dealt hereby are to define possible methods for the fabrication process, try-out some of these and finally implement the most highly accessible techniques in order to produce specimens for testing. To plan and optimize the production method CAD tools are used.

For the production of the specimens intensive experimental prototyping processes are followed. These often involve the use of power tools, commonly used in wood-working studios. Examples of further methods used for the fabrication of prototypes include casting, pressure forming, hand-lay-up composites etc. All processes are explained extensively in chapter (4.2). The prototyping experiences are consistently documented per case study with the aim to highlight the pros and cons and also how critical issues were resolved.

(4) Experimental structural testing

As there are no standard rules for testing paper-based components, some experimental methods are followed. Hereby an introduction to the subject is made. Extensive details about the testing conditions implemented and the processing of results are provided in paragraph 4.2, as guidelines for the case studies.

The main testing methods applied are the following:

- Experimental 4-point bending test (example available in paragraph 4.2).

A reference example is demonstrated in the project 'paper dome' of S. Ban (McQuaid, 2003, p. 79).

- Axial compression test (an example is available in page 212).

Two reference examples are demonstrated in the projects: 'paper house' and 'paper dome' of S. Ban (McQuaid, 2003, p. 77), where this testing method is used to examine the shear strength of screws that join the paper tube with the wooden insert.

Further testing methods attempted regard:

- Shear tests of basic joining techniques
- Experimental single point bending test of T-joint configurations
- Experimental torsion test (Nihat Kiziltoprak, 2019)

The testing device used for this purpose is a 'Zwick Roell' machine for material testing and the software used to monitor the testing data is 'testXpert II'.

The information extracted by the testing process is the following:

- Failure modes are identified
- F_{max}. vs. vertical deformation at the support (point where the force is applied)
- Homogeneity or scattering of results is observed (repetition of tests)

Further use of data is explained in paragraphs 4.2 and 4.3. There, the methods used to compare the different case studies and evaluate them are also presented in detail.

(5) Structural analysis with FEM simulations

Local FEM analysis is performed for selected case studies, with the aim to examine the distribution of stresses along the suggested assemblies. The design of the joints is modelled in detail in Rhinoceros software. 'ANSYS Workbench for structural analysis' is the software used as a tool for FEM.

In principle two different cases of Local FEM Analysis are examined:

- Multi-axial nodes, with the context described in paragraph 3.3.2.
- Linear assemblies, identical to the configurations used for the 'experimental 4-point bending test'.

The input provided is, in principle, as following:

- The essential material properties for each volume of the model are provided.
- The contact relation between the different volumes is defined.
- The mesh for all geometries is generated.
- The forces and support points are set.
- The calculation parameters are selected.

As output, the 'Maximum principle stresses' and the 'Total deformation' are calculated for the individual 3d-models.

Further information regarding the set-up of the simulations performed and the interpretation of the output is provided in paragraph 4.2 (guidelines).

(6) Further considerations

- Extra parameters

The impact of alternating climate effects on the performance of joints is not encountered. The importance of this aspect and the drawbacks on the structural behavior of paper products are known from relevant studies on paper-based materials. In a similar way, the fire resistance of the joints is not part of the investigations. Further research, including material testing and computational simulations is required to identify the impact of these effects.

- Constraints and uncertainties

The material properties of paper products cannot be generalized, as specific commercial products were selected and due to the variables within the production process, properties may vary. To explain this further, for example, the density, fiber length and the strength of the bonds between the fibers influence the material behavior significantly.

Literature

BAMP! (2017). Retrieved from https://www.tu-darmstadt.de/bauenmitpapier/startseite_1/index.en.jsp

Bank, L. C. (2016). Paperboard tubes in structural and construction engineering. *Elsevier*(Nonconventional and Vernacular Construction Materials. Characterisation, Properties and Applications), 453-480. Retrieved from https://www.sciencedirect.com/science/article/pii/B9780081000380000160?via%3Dihub

CEN. (2004). EN 1995-1-1 :2004+A 1. Eurocode 5: Design of timber structures - Part 1-1: General - Common rules and rules for buildings. In.

iggesund. (2020). The paperboard product. Retrieved from https://www2.iggesund.com/en/knowledge/knowledge-publications/paperboard-the-iggesund-way/the-paperboard-product/

McQuaid, M. (2003). *Shigeru Ban*. London N1 9PA: Phaidon Press Limited, Regent's Wharf All Saint's Street.

Nihat Kiziltoprak, E. K. (2019). *Load capacity testing method for non-conventional nodes joining linear structural paper components*. Paper presented at the ICSA.

Pohl, A. (2009). *Strengthened corrugated paper honeycomb for application in structural elements*. Retrieved from

Schütz, S. (2017). *Von der Faser zum Haus*: bauhaus ifex research series 1.

VDI, G. E. K. V. (2004). *Methodical selection of solid connections. Systematic, design catalogues, assistances for work*. Retrieved from

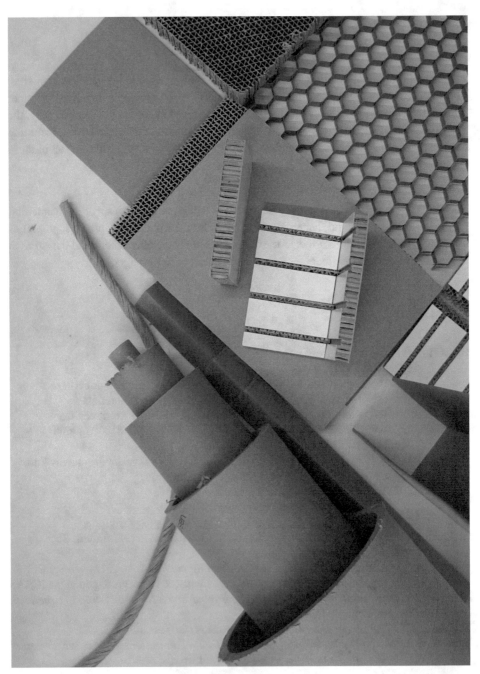

Figure 2.1 A collection of paper-based products (boards, profiles and rope) that are directly available in the market. These particular samples were fabricated in Germany.

2 Joining techniques - State of the art

Keywords: joining techniques, literature review, paper-based structures, paper tubes, honeycomb-boards, corrugated-boards, reference projects, structures, timber, bamboo, textiles, composites, steel, overview of joining principles

This chapter aims to develop a 'state of the art' for the subject of joining techniques for paper-based structures and to consider further possibilities influenced by common assembly methods and building practices. As building with paper is a novel subject, prior to the review, an introduction regarding the selection of realized design and construction projects, as an integral part of this review, is made. There, the focus on practical solutions and details developed within professional environments, and selection criteria for 'reference projects' are explained (paragraph 2.1.1). These projects are the context of implementation for the joining techniques that are analyzed later. Additionally, the special direction of this research on investigating joining techniques for beam structures composed of paper-based tubes, as presented in chapters (3) and (4) is justified, with the help of tables 2.1 and 2.2. In continuation, an 'overview of industrial paper-based products' (tables 2.3.1, 2.3.2 and 2.3.3) is provided, as well as additional information about the features that are expected to influence the potential use of these products for building purposes.

The review is mainly divided in two sections, one that regards the subject of 'joining techniques for paper-based products' (2.2) and one with focus on conventional building materials and assemblies (2.3) that could inspire further development within the previous section. The first section is divided in the three main areas of basic joining techniques for thin-sheet paper-products, techniques specialized on lightweight boards (such as corrugated, honeycomb and multi-layered assemblies) and techniques specialized on profiles (mostly paper-tubes and also L, U and square profiles). This structure creates the opportunity to discuss a wide range of solutions, from simple joining techniques found in everyday industrial products to designs developed especially for paper-based structures. In the second section the focus lies on common joining techniques, within conventional building methods, that show potential for implementation in paper-based structures. These regard materials such as timber, bamboo, textiles and composites that present a material structure with significant similarities to this of paper-based products. Additionally, solutions from steel structures are considered, as a category in which the aspect of joining techniques has been thoroughly investigated and has a wide spectrum of solutions to offer. The observations and outcomes from the review process are summarized in paragraph 2.4 that is concluded with a global overview of joining techniques that present good prospects for effective implementation in paper-based structures.

© The Author(s), under exclusive license to Springer Fachmedien Wiesbaden GmbH, part of Springer Nature 2022
E. Kanli, *Experimental Investigations on Joining Techniques for Paper Structures*, Mechanik, Werkstoffe und Konstruktion im Bauwesen 60, https://doi.org/10.1007/978-3-658-34501-3_2

2.1 Groundwork

This paragraph is divided in two sections. Paragraph 2.1.1 is meant to explain the idea behind the selection of reference projects, as part of the review process and further considerations in relation with this subject. Paragraph 2.1.2 presents a showcase of paper-based products and provides further insights, based on literature and research outcomes, focusing particularly on the subject of material behaviour.

Prior to the collections of references and paper-based products, targeted clarifications for the content of the whole review (sections 2.2 - 2.4) are made.

The review of joining principles and elaborate techniques includes a collection of pictures that present samples of materials, examples of joints, also within realized constructions. The majority of the assemblies presented are based on existing precedents, as these are referenced per case. When the content is produced within this research process, it is clarified as well.

The categorization of joining concepts, principles and techniques is inspired by guides, manuals and catalogs as stated in each sub-section (per material).

In section 2.2 a further distinction is made between the joining techniques for lightweight boards (2.2.3) and thin-walled profiles (2.2.4), due to the significant differences in the geometry and the matrix (layering) of the materials and consequently the particular application prospects in construction.

The distinction in the three main categories of 'material-closure', 'form-locking' and 'force-closure' appears in all sub-sections, when explaining the basic joining techniques. It is inspired by the design catalog for fixed connections, as presented in the guidelines for engineers regarding the 'methodical selection of solid connections', as an assisting tool for the engineering process (VDI, 2004, p. 26, catalog 1). Joining techniques that use more than one of these methods are identified as hybrid.

In section 2.3 the outlines of the review adjust as following. For each construction material discussed a short introduction is given at first, to describe its main characteristics and the relation identified with paper-based products. Then, a table with basic joining principles is presented, where a selection of solutions is visualized. These are further discussed, in continuation, within the categories of material-closure, force-closure and form-locking that are mentioned beforehand. There, in detail information is provided, accompanied with examples from realized assemblies and structures.

On this base, in section 2.4 the review of joining techniques for paper-based products and the applicability of further techniques from common construction materials are concluded. A total overview supplemented with explanatory tables summarize the outcomes of this process.

2.1.1 Collection of reference projects and studies

The aim of this paragraph is to demonstrate a collection of reference projects that are further discussed within the review (2.2) and explain the general grounds on which these were selected. A diverse range of studies on the behaviour and potential use of paper-based products in designs and structures were developed prior to this research. Here, a major distinction between structures composed of round paper-based profiles and those built with boards is made. As mentioned in the motivation (1.1), these profiles are the most widely available paper-based profiles and are popular for erecting temporary skeletal and truss-like structures. On the other hand, load-bearing assemblies built with boards refer mainly to plate or shell structures that present a different set of characteristics overall. The aspect of structural typologies is further discussed in paragraph 3.1.

Another general distinction between reference projects, design studies and studies with a greater focus on the material behaviour can be made. The aim is not to list all previous efforts, but to provide an overview of the references that influenced this research at the most. Hence, the focus lies on reference projects that contribute to envisioning the greater picture of how paper-based components are so far integrated within assemblies. In this spirit, the projects selected for further observation are mostly the ones where realistic boundary conditions were considered. During the review it was observed that for the structures integrated in public space, certain measures - precautions - were taken which often helped to advance the quality of the construction. To explain this better, there are no regulations about how to build with paper-based products and next to this, often paper-based designs aim to demonstrate a concept rather than a fully functional solution. Most of the selected structures were installed outdoors, temporarily and thus were prepared to withstand certain conditions, in terms of loads or climate alternations. Often the structures are not fully built with paper-based products. Still, it is particularly interesting to observe the combination with common construction materials and also how the advantages of the paper-based components are then used.

The tables 2.1 and 2.2 include a selection of realized paper-based structures that exhibit a variety of joining methods, considered as groundwork developed prior to this research. As mentioned beforehand, studies on the subjects of fully experimental prototypes and material characterization are not included in the following tables. However, these are referenced in the thesis (see list of literature per chapter). For example in the case of paper-tubes, further studies about the implementation of this product in construction are encountered, such as for example previous reviews about the applications of this component in construction (Lawrence C Bank, 2015). Similarly, in the case of boards there are numerous journals that deal with the material behaviour of the board itself (for different kinds of boards). Such issues are discussed in the following paragraph 2.1.2.

Table 2.1 A collection of primary projects, from literature, with reference to structures that use paper-based tubes as load-bearing members.

	Source	Nr	Name of Project	Location	Joining method(s) of interest
Book	(Jacobson, 2014)	1	Paper Log House	PH	Form-locking joints
		2		JPN, TR	Wooden joints (made with plates)
		3	Miao Miao Paper Nursery School	CN	Wooden Joints (made with beams)
		4	Hualin Temporary Elementary School		Wooden joints (two types, made either with plates or beams)
		5	Emergency shelter	LK (Haiti)	Wooden joints (made with plates)
		6	Emergency shelter for UNHCR	RWA	3d-printed joints
		7	Concert Hall	ITL	Wooden fundament joint (fixed in concrete with steel bars)
		8	Cardboard Cathedral Christchurch	NZ	Hinged metal joints
		9	Nomadic Museum	NY	Hinged metal joints
	(McQuaid, 2003)	10	Paper arch	JPN	Hybrid joints (made with metal and wood beams)
		11	'Boathouse'	FRA	Hybrid metal joints
		12	'Library of a poet'	JPN	Wooden nodes (made with beams)
		13	Paper House	JPN	Wooden fundament joints (made with beams and plates)
		14	Paper Church	JPN	
		15	Japan Pavilion EXPO 2000	DE	Lap joints with ties
		16	Paper arch, modern art museum	US (NY)	
	(Miyake, 2009)	17	'Paper Arbor'	JPN	Columns capped with wooden compression ring
		18	Odawara East Gate	JPN	Bolted cubical-shaped steel node
		19	Paper Church – Kobe	JPN	Wooden fundament joints
		20	Paper studio	JPN, FRA	Massive timber Joints

		21	Biennale Pavilion	SG	Concrete nodes
	(TU-Delft, 2008)	22	Cardboard Dome	NL	Steel multi-axial nodes
		23	Westborough School	UK	Pre-stressed tubes
Journal	(Preston, 2011)	24	Paper-tube arches	US	Bolted lap joints
Web-page	(Cottrell-Vermeulen-architecture, 2001)	25	Westborough Primary School	UK	Steel joints (connections to beam and fundament)

Table 2.2 A collection of primary projects, from literature, with reference to structures built with lightweight paper-based boards.

	Source	Nr	Name of Project	Location	Joining method(s) of interest
Book	(McQuaid, 2003)	1	Nemunoki Children's Art Museum	JPN	Metal connectors between the cardboard ribs
	(Schütz, 2017)	2	Prototype of a cubical unit	DE	Lap laminated joints Interlock - joints with flaps
	(Miyake, 2009)	3	Paper Shelter	JPN	Basic assembly of honeycomb boards with use of tape
	(TU-Delft, 2008)	4	Apeldoorn Temporary theater, ABT engineers	NL	Paper nodes reinforced with wood, between the cardboard ribs (arch-shaped structure)
Web-page	(Cottrell-Vermeulen-architecture, 2001)	5	Westborough Primary School	UK	Multi-layered cladding panels
	(Fiction-Factory)	6	Wikkelhouse	NL	Laminates Pressure rods
	(Latka, 2016)	7	House of Cards	PL	Nails and screws
	(CelluTex)	8	Interior design products	NL	Laminates, stitches and screws

Based on the review process, the product of paper-based round profiles is very popular for the development of temporary structures. The main reasons for this are related to the manufacturing process that allows for the production of profiles with different geometrical characteristics and qualities in terms of strength (based on the paper-sheets used and the wall thickness). In addition to this, the round shaped profiles present structural advantages comparing to profiles with sharp corners and in combination with the solid wall, the structural capacity, particularly against axial compression, reaches promising performances. Specific information about the structural performance of the paper-tubes is provided in paragraph 3.1.2. On the other hand, multi-layered paper-based composite boards require the development of modern equipment to achieve effective results in manufacturing and complex studies to examine the behaviour of the finished components. These issues are further discussed in the following paragraph.

2.1.2 A brief introduction to paper-based products

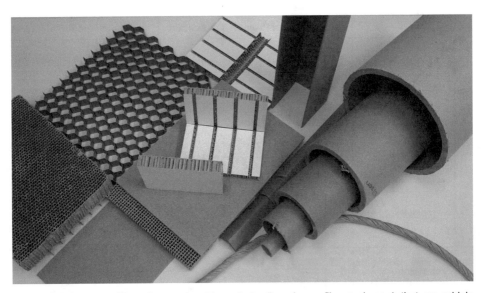

Figure 2.2 A composition of paper-based products (boards, profiles and rope) that are widely available by the paper-industry. The samples demonstrated in this picture are fabricated in Germany, a country known for the high-standards in the manufacturing of paper and paper-products.

Paper products are globally used in many sectors, such as these of hygiene products, office supplies, shipping, draft prototypes, interior design and temporary small-scale installations. Hence, there is a broad variety of paper products with different qualities that satisfy the aforementioned applications. The production process is constantly being optimized (ETSAP, 2015).

The tables 2.3.1, 2.3.2 and 2.3.3 provide an overview of paper products that are more relevant to the subject of construction. Even though all these products have the same origins they present a different material matrix, depending on the production process, the layering and simple geometrical factors that are also related to the application. Thus, in table 2.3, all products are clustered in three different groups.

Looking into the production process, a distinction between two main fields can be directly identified. The first one is paper mills and the second one is factories that post-process the material produced by the mills, to form specialized products such as lightweight boards and profiles. So, the first stage of the production ends with a roll of paper-sheet that presents pretty much the behavior commonly known, with advantages for the tensile strength of the sheet in the direction of production. However, the post-processing methods often lead to end-products that present significantly different behavior from a structural perspective. This aspect is very crucial for the structural application. Therefore, in table 2.3, for each product, the main forming process, range of wall-thickness and weight per unit are provided.

Table 2.3.1 Basic paper products and boards

Material	Forming Process	Wall Thickness (min, max - mm)	Grammage (g/m²)
Paper	Pressing	(0.07, 0.18)	8 – 150
Carton	Pressing	(> 0.2)	170 – 450
Paperboard	Pressing	(> 0.3)	> 600
Corrugated paper (Single wall)	Curving and gluing	(0.25, 5)	320 – 775
Corrugated paper (Double wall)	Curving and gluing	(0.5, 10)	525 – 1250
Corrugated paper (Triple wall)	Curving and gluing	(0.75, 15)	730 – 1400
90° corrugated board	Curving, gluing, lamination	(5, 82)	Varies
Honeycomb board	Adhesive bonding	(5, 82)	Varies

Table 2.3.2 Paper-based profiles

Material	Forming Process	Wall Thickness (min, max - mm)	Grammage (g/m²)
◯ Circular profile	Winding	(> 0.3), customized	Varies
▢ Rectangular profile	Winding	(> 0.3), customized	Varies
▬ Flat Profile	Lamination	(> 0.3), customized	Varies
L Profile	Lamination and Bending	(> 1), customized	Varies
U Profile	Lamination and Bending	(> 1), customized	Varies

Table 2.3.3 Other types of paper or pulp-based products

Material	Forming Process	Wall Thickness (min, max - mm)	Grammage (g/m²)
Cellulose foam	Casting	varies	Varies
Pressed pulp	High-pressure molding	(1, 10)	Varies
Casted Pulp	Casting	varies	Varies
3d printed cellulose	3d Printing	varies	Varies
Paper yarn	Twisting	(0.07, 0.18)	Varies

The data for the grammage provided in table 2.3.1 are based on material guides (Iggesund-paperboard).

For the products presented in table 2.3.2, the grammage cannot be specified, as there is no official range and the specifics depend per supplier.

In a similar way, in table 2.3.3, the grammage of most products varies significantly, based on the composition of the pulp-based mixture. Most of these are still in the development or research phase (Halonen, 2012) and their actual implementation is still at a preliminary stage (except for the paper yarn for which commercial products are accessible). In principle, especially for industrial applications, the binders used and possible combination with other elements to reinforce the matrix play a significant role in the weight.

A variety of fundamental forming processes for paper-based products is shown in fig. 2.3 Various manuals describe extensively the forming process of paper-sheets, from pulp to thin sheets (Iggesund-paperboard, p. 32). There interesting information about the material composition and the integration of adhesives on the materials' surface, to improve strength is presented. Lamination is a fundamental method used to form thicker products. It is also crucial in the production of corrugated- and honeycomb- boards. A different approach regards post-processing of paper-based products to change their shape by deforming them with high-pressure. In this process, the material is often steamed in the start and dried at the end. For example, to form L and U profiles, flat paper-board profiles are bent. In a different way, complete molds can be used to form components with different shapes, using high-pressure and multiple layers of paper-based sheets. Casting pulp is another known forming technique that is commonly used to craft art objects. In principle, the greater the volume and fiber content, the more challenging it is to control the quality of the final product and prevent shrinking or swelling, as climate alternation have a greater impact on the humidity content.

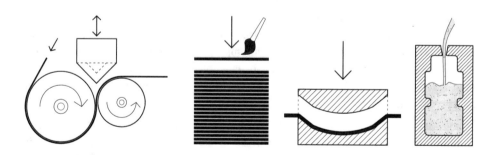

Figure 2.3 Manufacturing processes from left to right: extrusion, lamination, high-pressure forming, casting

Furthermore, there is a great variety of sophisticated machines and equipment available for the manufacturing of paper-products. Actually, the pulp and paper industry reportedly is one of the most energy-intensive sectors of the European union an issue that is targeted for improvements in the direct future (SETIS).

2.1.2.1 Features & observations derived from existing studies

In the field of design and construction, little is known about the material composition and structural behavior of paper products. In this research, these aspects are particularly important, as individual characteristics eventually lead to significant effects on the performance of potential joining techniques.

A selection of examples from existing researches on the aforementioned aspects that add up to the global picture of what is known so far are described below. The focus is both on understanding the material behavior of paper products and also looking into effective testing methods, as within this research experiments are part of the investigation process.

The products discussed are these of paper, paperboard, corrugated and honeycomb boards, in an effort to present the similarities and differences between these materials. For the case of profiles made of paper or paperboard, prior findings deal mainly with the round profiles, a product that is discussed extensively in paragraph 3.1.2.

Material characterization

In general, up until now, the characterization of the structural behavior has proven to be a complex task, as it differs significantly per product. Keeping in mind that paper-products are originally meant for general use, building with paper is a subject that introduces an entirely new possibility and requires intensive studies in this direction.

Taking paperboard as an example, the kind of pulp used (bleached or unbleached chemical pulp, mechanical pulp, secondary fiber pulp) can lead to products with enormously different performances. Next to this, the material structure, composed of relatively short fibers, imposes significant challenges. In addition to this, the surface treatment (coated or uncoated) also plays an important role, especially for the durability of the product. At the same time, when, for example, paper-products are used to form multi-layered composite panels, such details are crucial, to ensure a functional assembly and prevent delamination.

An introduction to this topic is made within research developed by TU Delft, aiming to identify the fundamental material properties of cardboard, characterize its behavior (material modelling) and examine a number of case studies for composite beams through experimental structural tests (Schönwälder, 2008).

- Anisotropy

According to studies, paper demonstrates highly anisotropic and nonlinear mechanical behavior that require further studying when considering the combined effects of moisture and temperature on its mechanical response (Yujun Li, 2017).

A representative example, within this area, is shown in a study that analyzes the structural response of paper and cardboard with a computational model that incorporates the creep and tearing rate dependence, highlighting the effects of time dependent behavior on crack growth. (Schönwälder, 2016).

For failure in sliding mode, the importance of having more fibers oriented in the direction where cracks are expected to occur, rather than transversely to it, is indicated within a study focused on delamination of the paperboards interface (Y. Li, 2016).

- Durability vs. climate

The hygroscopic behavior of paper and its products is a parameter that has been under investigation (Schönwälder). An elaborate study about 'the combined effects of moisture and temperature on the mechanical response of paper' (Linvill, 2014) provides interesting insights, in an effort to generate data for the development of a 3-dimensional material model for paper. Tensile tests are used as the basis for the experimental testing process, in order to examine certain material and failure properties. The effects of moisture and temperature are studied individually. The conclusions on the failure modes indicate coupled moisture and temperature effects for maximum stress and uncoupled effects for strain at break and approximate plastic strain, whereas tensile energy absorption appeared to be significantly affected by increase in temperature only. This way, parallel increase in temperature and moisture would decrease the maximum stress of the specimen at failure. At the same time, the strain at break is expected to increase with higher moisture and decrease with higher temperature, whereas the durability of paper appears to decrease when exposed to higher temperatures. Keeping in mind that upscaling of the specimens leads to more challenging conditions, as the weak zones increase, this problematic would be very interesting to study for further paper products, also in dimensions with potential for design or structural applications.

- Internal bond strength

In continuation of the discussion over the kind of pulp, the chemistry used in the forming process and potential addition of adhesives, influence significantly the bond strength, between the fibers. This way, the internal bond strengths differ per case. Different methods can be used to measure and identify this particular behavior. A study has indicated the z-directional tensile test (direction of material-thickness) as the most suitable to identify pros and cons between different products. The same study concludes that chemical pulps have better bonding abilities than chemi-thermomechanical and thermomechanical pulps, but the second ones present the highest specific bond strength. (Koubaa, 1995). According to the results, higher density corresponds in principle with increase in fiber-to-fiber bonding. In a different way, lower basic weight seems to allow the adhesive to create a reinforcing effect. Moreover, the study suggests that the bond strength perpendicularly to the surface of the material decreases with increase in basis weight or in sheet thickness, as in these cases the number of weak zones increase. In other words, paperboard with lower basic weight is predicted to present lower z-directional strength. Next to this, when the thickness of a board increases the z-directional strength of the internal bonds is expected to decrease. These findings are very useful when, for example, considering creasing or folding the paperboard or manufacturing multi-layered panels.

Investigations on lightweight hollow products

The context becomes more complex when it comes to sandwich plates, such as corrugated and honeycomb boards, due to structural instability of these in certain directions. For example, honeycomb boards are susceptible to delamination along the inner face of the side liners due to shearing. On the other hand, corrugated boards are relatively compressible, with soft edges parallel to the direction of the flute. These characteristics challenge the standard methods for material testing.

Several attempts to study the material behavior of boards have been done. These start, for example, with efforts to characterize the behavior of single plates and continue with studies about more complex (multi-layered) components.

In this spirit, within the process of testing plates of corrugated board, researchers often look into particular fixation methods for the specimens. An example refers to an edge-compression fixture for examining the effects of buckling (Hahn, 1992). The aim then is on proving that the specimens fail due to buckling and not due to failures occurring within the region of the fixation.

Encouraging findings are presented in a relevant study that analyzes the behavior of adhesive joints in double-lap corrugated paper under shear loading, with structural tests in comparison with a simplified computational model (Conde, 2012) .

More complex effects, such as twisting (Perez, 2013) and torsion, have been in the focus of researchers, in efforts to generate data and conclusions in the direction of enriching the methods and tools for structural assessment. Further studies that could be beneficial for researchers who investigate the potential of paper products for structural applications can be found in the sector of packaging for goods (Fadiji, 2016).

Though the subject of structural analysis is not in the main focus of this research, the findings reviewed in that area indicate that the lack of standardization imposes great challenges for engineers in the field of construction, as well as the lack of tools for direct evaluation of components and in continuation assemblies. Hence, even though there is rich input about the manufacturing process and characteristics of paper, further research is required to understand how its products would respond if used as structural components.

2.2 Joining techniques for paper-based assemblies

2.2.1 The architecture of the review

While structuring the review of joining techniques, for paper-based assemblies and structures, two main factors were considered. The one regards the perspective and approaches appearing in previous researches. The other one is related to the greater vision of this research, meaning potential construction methods and applications. These topics are briefly discussed below, as an introduction to the review process, to address considerations made during the preliminary phase of this research.

2.2.1.1 Joining techniques for paper-based assemblies prior to this research

As it is mentioned in the motivation (paragraph 1.1), the subject of joining techniques for paper-based structures has not been yet in the main focus on its own. To present this situation concretely, two specific examples of research precedents that attempt to briefly frame potential joining techniques are given below.

An example is presented in the dissertation of Özlem Ayan, with general focus on the implementation of lightweight paperboards in building applications (Ayan, 2009, p. 166), in the form of multi-layered panels, used as wall components. There, the four main characteristics considered important for the performance of the joining techniques are namely: 'flexibility', 'location classification' and 'functional classification'. This abstract categorization follows the requirements that steam from the researched wall-component. It basically helps to sketch a global picture about the functionality of joints that should be studied when someone would aim to bring the wall-component into practice.

In a different way, the dissertation of Jerzy Latka in the field of architecture that is more related to the conceptual design of cardboard structures in general, discusses the aspect of joining techniques, focusing mainly on the assembly method. Following an extensive review of existing projects and studies developed in the context of student assignments, seven main practices are distinguished (Latka, 2017, p. 259) for the aspect of joining techniques. These are namely: 'lamination, screw and bolt, bracing, post-tensioned elements, interlocking, folding, clipping and tiding, connections to the ground and the aspect of impregnation'. As part of the observations, it is addressed that adhesive joints appear to be the most effective, whereas flexible joints the least.

This research adopts a new approach for the review and categorization of the joints, as explained in detail in the following paragraph.

2.2.1.2 Framing the review of joining principles & techniques

The conceptual construction details presented in figures 2.4 and 2.5 demonstrate the global vision, when thinking about building with paper. These details support the two cases of building with load-bearing composite plates or designing a load-bearing frame using paper-based round profiles, the most effective profile for the time being.

In the vertical sections shown in figures 2.6 and 2.7, critical joints between construction elements are pointed out. As up to this point there is no proven solution, the first priority, in each case, would be to resolve the primary joints. Based on the review of realized temporary structures, the use of other materials, such as wood as reinforcement, at these critical areas, is a common strategy. In the review, the discussion of these issues is in the main focus.

A key-point in the search for functional joining principles is the fact that the majority of existing solutions have their origins either within the production of common paper products or in common types of construction (timber etc.). Under this consideration the review is divided in two main sections, one devoted to paper-based assemblies and a second section where the relation of the previous ones with conventional materials and joints is investigated.

As explained in the motivation, joints are clustered following the principle fixation types material-closure, force-closure, form-locking and are examined separately for three different categories of paper-based products. Hence, a distinction between basic joining techniques (2.2.2) and specialized solutions for lightweight-boards (2.2.3) and paperboard-profiles (2.2.4) is made. Eventually, for each category, a showcase of techniques is presented, either in the form of diagrammatic overviews, or with construction details that include the most relevant joining methods.

When examining the influences from common construction materials the focus is on addressing assemblies that could potentially be built with paper-based components. Representative examples for this are traditional timber frames (fig. 2.58), bamboo frames (paragraph 2.3.2), versatile connectors from textile structures (table 2.9) etc.

All conclusions are incorporated in a total overview (table 2.12) that is supplemented with observations on the pros and cons of the individual suggestions.

A preliminary observation about the direction of this research, following also issues related to the material behaviour and the current manufacturing potential, discussed in paragraphs 2.1.2 and 2.1.2.1, is that the design of load-bearing composite plates is a rather challenging subject on its own, especially to ensure uniform quality. As in this research the main target is the joints, looking at the conceptual details (fig. 2.4 – 2.7), skeletal structures composed of round profiles provide better grounds for the elaboration of case studies. This discussion is further analysed in paragraph 3.1.

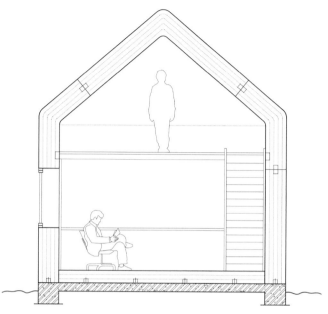

Figure 2.4 Conceptual construction detail, vertical section: plate-structure composed of multi-layered load-bearing walls, made with paper-based lightweight boards.

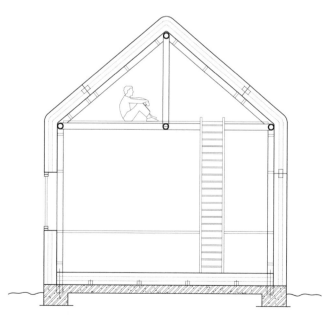

Figure 2.5 Conceptual construction detail, vertical section: skeletal structure, composed of round paper-based tubes and cladded with multi-layered paper-based lightweight boards.

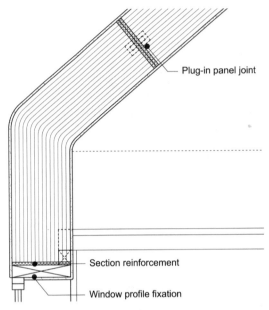

Figure 2.6 Conceptual construction detail, vertical section: Fundamental junctions in the case presented in fig. 2.4.

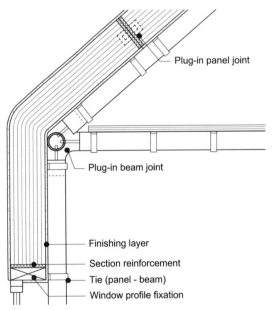

Figure 2.7 Conceptual construction detail, vertical section: Fundamental junctions in the case presented in fig. 2.5.

2.2.2 Basic joining techniques

Below the most common practices used to join flat solid paper materials are presented. Some of them are commonly used in paper-forming processes, production of packaging, book-binding, arts and crafts and textile products (2.3.3).

Table 2.4 Principle joining techniques for paper-based products and particularly thin solid boards.

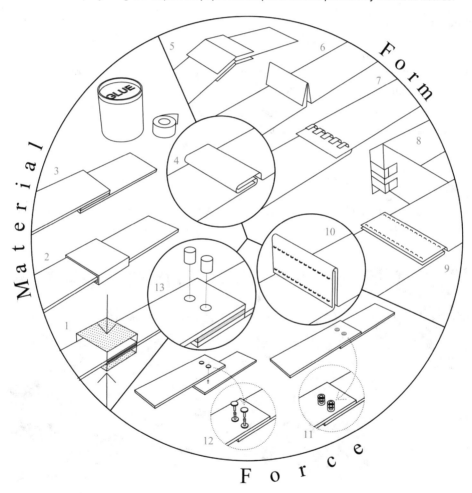

Material-closure

1. Couching
2. Tape
3. Adhesion
4. Adhesion + folding
5. Crease and overlap

Form-locking

6. Fold together
7. Interlock 0° (flexible)
8. Interlock 90° (rigid)
9. Stitch (flexible)
10. Staple (rigid)

Force-closure

11. Pressure ring
12. Rivet
13. Dowel & adhesion

Material-closure

Following technical reports and articles from the paper industry, the most common adhesives used in paper production are the following:

- Starch
- Dextrin (also starch based)
- Water-glass

Starch and dextrin are known as low cost adhesives that bond well with paper products. Starch might be the most popular eco-friendly binder in the paper industry. On the other hand, dextrin, presents lower green strength, lower viscosity and thus higher workability, so it is sometimes preferred for certain machinery. Both adhesives are known to perform poorly in conditions of high humidity and moisture in contrast to water-glass.

Moreover, adhesives that are commonly used for the assembly of strong packaging products and small-scale paper structures are:

- PU glue
- PVA glue

In principle, PU glue is absorbed less through the surface of application comparing to PVA. The bond strength is comparable, with PVA having a slight advantage. PVA glue is used exclusively for wood products, whereas PU glue is suitable for a wide variety of materials.

Figure 2.8 Prototype of a foldable origami-design made by a group of students, 2014, Buck Lab studio (Fall 2014, TU Delft, direct supervisor: Jerzy Latka, head of studio: Marcel Bilow). The triangulated shell is made with corrugated paperboard pieces and flexible joints made with pieces of textile and tape, a solution that is good enough to demonstrate the design concept but is not meant for real use.

To close-up the brief reference on adhesives, an interesting example of a house in which the majority of surfaces were formed with laminated newspaper, built in 1922 already, by Elis F. Stenman, in Massachusetts (US), shows the early fascination for re-using paper-products in a creative manner.

Some of the adhesives mentioned above are also used in tape products that are often preferred either for rapid (0 drying time) or for reversible assemblies. Single-sided tape helps to form a sealed bond, whereas double-sided tape can provide a steady lap joint.

An experiment on adhesive joints

The purpose of this study is to get a glimpse of the failure mechanism of laminated joints for paperboard when subjected to shear and examine the behavior of these depending on the main direction of the fibers in the board. The adhesive used is 'Ponal' glue for wood. Two different configurations of specimens were tried: a simple lap joint in the middle with double-laminated ends (fig. 2.9, left), to prevent eccentricity and a double-lap joint in the middle with single layers of board on the sides (fig. 2.9, right). Both configurations were found to be functional. The first configuration was selected due to the simplicity of the assembly that helps to minimize errors related to potential imperfections in the lamination process.

 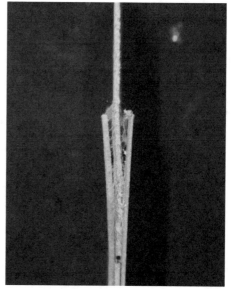

Figure 2.9 Series of shear tests with specimens of laminated paperboard, performed in this research, at the laboratory of ISM+D (02.2019), with the aim to get a glimpse on the limits of the laminate, confirm the suspected failure mode (fracture due to delamination of the top-layer of the paperboard) and examine the relation between the primary fiber orientation and the performance of the laminate.

The results showed that the ultimate strength of all specimens within the elastic region is comparable (about 0.8 kN approximately, based on graphs 2.1 and 2.2). The specimens fail due to delamination within the paperboard and close to the surface that is in contact with the adhesive, as it was expected. The specimens laminated along the fiber direction demonstrate linear curves within the elastic region, resulting to abrupt failure. On the other hand, the specimens that were laminated perpendicularly to the primary fiber-direction presented higher durability within the range of plastic deformation. Based on these outcomes the importance of the properties of the faces of the boards is highlighted and the idea of cross-laminated paperboard is encouraged.

Graph 2.1 Experimental shear test series performed with laminated specimens of kraft paperboard (dimensions of laminated area 5 x 10 cm). The lamination of the plates is made perpendicularly to the primary axis for the fiber orientation.

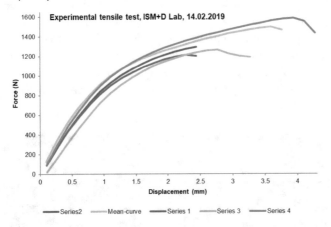

Graph 2.2 Experimental shear test series similar to these in graph 2.1, with the only difference that the lamination of the plates is made parallel to the primary axis for the fiber orientation.

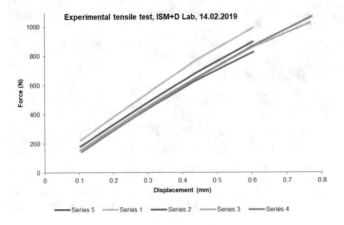

Force closure

Stitching is commonly used in book-binding. Some design products also use stitching, either to connect subsequent sheets or to secure and tension a finishing surface of paper around the perimeter of the design object. An example is the office desk made of paper materials designed and produced by (CelluTex) (AMS, NL).

Staples are preferred to connect thicker surfaces, where stitching is not possible, or to stiffen a lap joint - as staples decrease the flexibility of the assembly, comparing to stitches. The maximum depth that can be reached is about 14mm. A series of experiments, in the same spirit (same material, geometry of specimen and testing method) as these shown in fig. 2.9 lead to the following observations. At first, direct comparison between a laminated and a stapled joint is not possible, as the one joint is uniform and the other punctual. Still, the laminated specimens presented significantly higher ultimate strength than the stapled ones. The failure occurred due to a combination of tearing on the paperboard and deformation of the staples. The advantage comparing to lamination is that a joint composed of multiple staples tends to stay in place, even after failing and does not fully detach. In addition to this, based on the experiments, applying the connectors with direction perpendicular to the fiber orientation proved to be important. This was expected as then the material structure presents higher resistance to tearing. Further experimentation is required to develop a proper stapled joint and present exact values, as it is important to define the exact function, pattern of staples, material thickness etc.

Figure 2.10 Examples of common joining techniques for general use that are directly applicable on thin paper-based products. Left: Pressure rings, a solution known from textile materials, center: staples – a solution known from office equipment, as well as a basic assembly method for soft-wood boards, right: bolts – a common method of fixing plates. (The two last pictures were captured during workshops that took place during the interdisciplinary courses on façade technology, winter semester, 2017-2018. For further details see paragraph 6.4.2.1 in the appendix.)

To continue with further joining principles, fasteners that are used in textile products (paragraph 2.3.3.1) are also applied in paper products. Pressure rings (fig. 2.10) are used to create protected holes, so that a rope can be threaded. Rivets are used for lap joints, for example to create a lock at the opening flap of a box. Occasionally, velcro is used in the same way. Furthermore, office supplies include elements such as paper-clamps that are often used in rapid prototyping for architectural models and art installations.

Form-locking

Considering thin paper-products, this category is mostly limited to folding techniques, occasionally in combination with adhesion or stitching. Deployable products, such as origami models or boxes, use the advantages of creasing paper to form a stable structure. Flaps that are interlocked between the folded edges increase stability. For the same reason, occasionally, combination with adhesion is used. Moreover, local reinforcement along the edge of the joint is also helpful, either to increase the stiffness, or in the opposite way, to ensure the durability of a structure that needs to be deployed multiple times. In the first case a stiff paper product is used -narrow L profiles of paperboard are very popular for edge protection-, whereas in the second case a flexible one. Flexibility can either be related directly to the material behavior or it can be created through the design of the reinforcement. Textiles can also be used in this case (2.3.3.1, table 2.9).

Vulcanized paper combines strength and lightness and is used for high-performance packaging products. Its downside is that it cannot be recycled.

The principle of 'puzzle interlocks' is also worth-mentioning as it could be used to assemble flat surfaces in a larger scale.

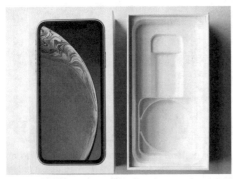

Figure 2.11 Treatments such as creasing and folding are traditionally used to create light and functional 3-dimensional objects with paper-based products. (Left) Packaging product where the outer protective layer is made with creased paperboard, (right) the strategy of subtracting material to create a flexible edge is also popular for foldable assemblies. The picture shows a laser-cut MDF surface, still the same principle applies for paperboard, in this case the durability of the edge is expected to be lower.

2.2.3 Joining techniques specialized on boards

Figure 2.12 A collection of paper-based boards with corrugated or honeycomb core. In the case of corrugated paper, there are a few different possibilities for the design of the flute-structure, whereas in honeycomb products, the design of the honeycomb shells may vary significantly in different products (design of perimeter and size).

The most prevalent joining principles for lightweight paper-based boards are demonstrated in table 2.5. In the following paragraphs, the techniques that belong to the categories of 'form-locking' and 'force-closure' are described. Regarding the category 'material-closure', the same information presented in paragraph 2.2.2 'Basic joining techniques' applies. Lamination is often used to create multi-layered boards (for example triple-wall corrugated boards) or even composite panels with high thickness. In continuation three reference projects are discussed, with focus on the joints and the assembly method.

Form-locking

Most form-locking principles are inspired by wood-working. For some of the form-locking principles shown in table 2.5 prototyping experiments were realized as part of the educational module 'Experimental Façade Technology 1' (appx., 6.4.2.1). Due to the fact that the core of the boards is hollow, such joints are much weaker than in timber construction, as friction is lacking and the edges become softer through time. The integration of timber blocks might enhance the performance. Bending the boards is often a method used to avoid an excessive number of joints within an assembly and also increase the stiffness of the structural element. Thus, it is encountered in the joining techniques in order to introduce the possibility of treating the material instead of joining two individual elements along the common edge. An interesting study on producing and forming lightweight beams with honeycomb board that also deals with the assembly of these elements in a tiny space is presented in the book 'Von der Faser zum Haus' (Schütz, 2017, p. 157, beam-to-beam-joint).

Figure 2.13 Avoiding joints by designing continuous folded edges, realized either by bending the board while steaming (left) or cutting angled edges within the board itself (right).

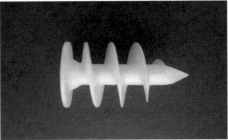

Figure 2.14 (left) 3-dimensional interlocks (this image was captured during the interdisciplinary educational module of IPBU, summer semester, 2018-2019, for further information see paragraph 6.4.2.3 appx.) and (right) plastic screws with wide spiral, are both common practices for the assembly of light boards.

Force-closure

In recent times a few suitable fasteners have been developed. One of these is screws with extra wide spiral for optimal anchoring, often used in paper-based palettes made with honeycomb boards. In this case a plastic version is used. Corrugated board though is already too dense and thus the plastic screw might only function with pre-made holes. Another type is steel cross-dowels, commonly used in lightweight furniture.

Table 2.5 Principle joining techniques for paper-based products and particularly for thick lightweight boards, such as multi-layered composite plates composed of laminated honeycomb or corrugated and solid boards. The suitability of the different joining techniques varies, depending on the matrix of the finished composite plate.

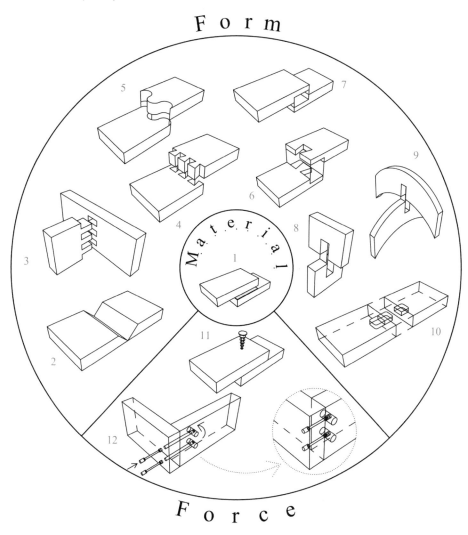

Material-closure

1. Lamination
2. Cut and bend
3. Dovetail joint 90°
4, 5. Dovetail joint 0°

Form-locking

6. Dovetail lap interlock
7. Simple lap form-lock
8, 9. Cross-interlock (ribs)

Force-closure

10. Biscuit
11. Screw with wide spiral
12. Cross-dowel

2.2.3.1 Construction details

In this section three reference projects are used as examples of implementing paper-based products in construction. These projects are developed in professional environments and provide insights about the design of the joints.

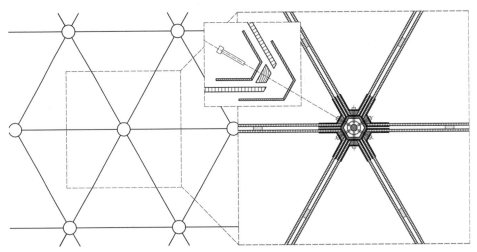

Figure 2.15 Detail drawn based on the project: 'Nemunoki Children's Museum' (JPN), 1999, S. Ban. The focus is on the composite node, that includes honeycomb plates fixed together on a steel ring with corner profiles and bolts.

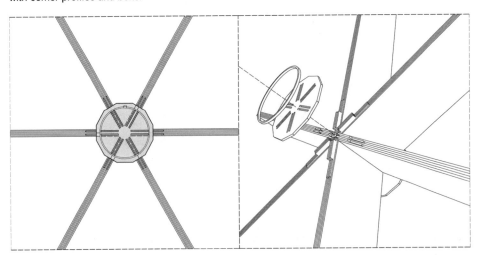

Figure 2.16 Details drawn based on the project: Apeldoorn Temporary theater (NL), 1992, ABT consulting engineers. The lamella structure is mostly built with paperboard. The nodes are made of laminated paperboard. Side caps prevent axial and rotational displacements and metal stripes fixed on top of these, along the perimeter, secure the nodes under pressure.

Figure 2.17 Project: Apeldoorn Temporary theater (NL), 1992, ABT consulting engineers. Inside view

Figure 2.18 Project: Apeldoorn Temporary theater (NL), 1992, ABT consulting engineers. Testing the building system.

 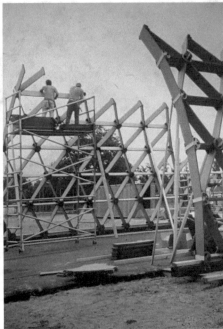

Figure 2.19 Project: Apeldoorn Temporary theater (NL), 1992, ABT consulting engineers. Component testing (left) and the erection process on the site (right).

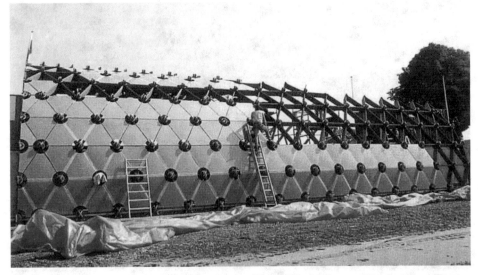

Figure 2.20 Project: Apeldoorn Temporary theater (NL), 1992, ABT consulting engineers. Placing the cladding on top of the lamella-structure.

Two main reference projects that share a common topic, the design of a node for a ribbon structure are presented hereby. Each of them presents a different approach on how to realize a lamella structure with paper-based materials, as a result of the requirements regarding the durability of the installation.

In Nemunoki Children's Museum, JPN (McQuaid, 2003, p. 56), the triangulated flat roof structure is composed of ribs made of honeycomb-board. Each rib is formed with a double-wall of honeycomb board with spacer in-between. The connections are intensively reinforced with steel elements (see details in fig. 2.15). The honeycomb boards are clamped between steel edges that are bolted on the central steel part of the node.

On the other hand, in the case of the arch-shaped 'Apeldoorn Cardboard Theater' (ABT engineering, NL) that was erected for a short time, few other materials interfere. The ribs are made of paperboard and are reinforced with wood at the area of the node (see details in fig. 2.16). Paper-board caps placed at the top and bottom of the node secure the ribs together and ties apply pressure.

A more modern reference project of a mobile house, a product named the 'Wikkelhouse' (fig. 2.21-23), presents a very interesting approach on how paper materials can be integrated in a long-lasting structure. In this case, multi-layered corrugated paper is used as thermal insulation. Layers of timber placed on the sides and middle protect the paper-core by carrying the loads. Next to this, the paper-core is fully protected and no holes are made on it, to guarantee good performance and increase durability.

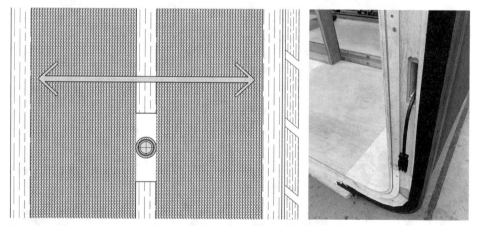

Figure 2.21 The product 'Wikkelhouse', made by the 'Fiction Factory', a company based in Amsterdam (NL). This company takes care of the design, manufacturing and installation of the house. It is composed of modules with a width of about 1.2 m. The wall is built with a continuous layer of wood on the inside and a timber frame on the outside, filled with a core made of corrugated paper. The modules are connected with steel bars that hold the whole unit under pressure (see detail on the left).

Figure 2.22 The product 'Wikkelhouse', made by the 'Fiction Factory' (Amsterdam, NL). The process of laminating the inner layer of wood on the corrugated-paper-core (image captured on 12.2019).

Figure 2.23 The product 'Wikkelhouse', made by the 'Fiction Factory' (Amsterdam, NL). There are two kinds of modules. The ones made simply with the paper-core as described beforehand (right) and those that include services such as ventilation openings, windows etc. that are composed of a more detailed timber sub-frame filled with thermal insulation (left).

2.2.4 Joining techniques specialized on profiles

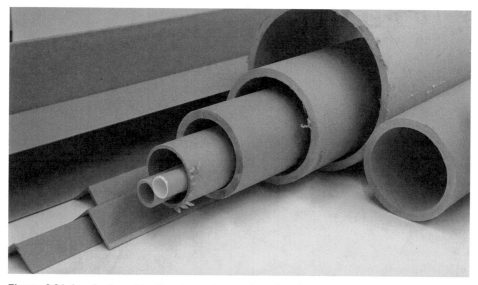

Figure 2.24 A collection of L, U and round paper-based profiles that are readily available in the market. Square profiles are less common, especially with a wall thickness that is higher than a few millimeters (4mm approx.), due to complications in the production process.

The majority of existing solutions regards round profiles, as there are numerous examples of implementation in temporary structures. Reference projects are the main source of information. The experiences of engineering offices that contributed in the development of these projects and a selection of books that present the efforts made to realize such experimental structures provide important input for this research (see table 2.6). On these grounds, conceptual construction details that aim to highlight the most important elements for a number of different assemblies are presented hereby.

2.2.4.1 Joining techniques for paper-based tubes

As an introduction, a collection of principle joining techniques for paper-tubes is provided in table 2.6, that summarizes the findings of the review (techniques 1-9) and efforts made within this research to identify further potential solutions (techniques 10 - 12). In continuation, this paragraph is divided in two sections that examine a range of different possibilities. The first one describes basic joining techniques that fall under the categories of force-closure and form-locking. In the second section hybrid techniques are presented that combine a number of different joining techniques together. There a further division takes place, based on the main material used for the joint, as following: light timber nodes, massive wooden nodes, metal nodes, 3d printed nodes, concrete nodes.

Table 2.6 Principle joining techniques for paper-based round profiles.

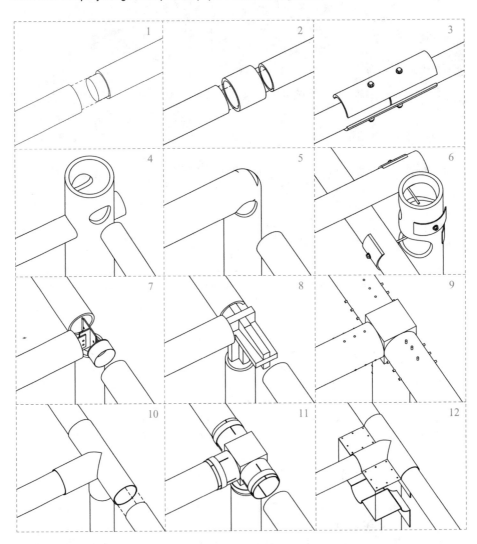

Linear joints	Form-locking joints	Plug-in joints	Sleeve joints
1. Inner-plug	4. Full-interlock	7. Metal plug	10. Simple form-fitting
2. Outer-plug	5. Tie-lock	8. Light cross plug	11. Intermediate connector
3. Pressure-profiles	6. Pressure-lock	9. Full contact plug	12. Clamp (pressure-plates)

Basic techniques

Force-closure

- Tie wraps

Simple weaving techniques, based on joining principles used in the field of bamboo structures (see 2.3.2) become also popular in paper-tube structures (fig. 2.26).

In the detail for the 'paper arch', shown in fig. 2.28, tie wraps secure lap joints between bent paper-tubes, in combination with steel threaded fasteners applied with bolts and pressure-plates. Tensile cables are fixed diagonally between the threaded rods. This system helps to maintain the initial distances between the layers of tubes and prevent them from collapsing into each other with time.

Tie wraps are used in a similar manner in the 'Japan Pavilion Project', realized in Hannover (DE) (McQuaid, 2003, p. 60). This structure is of a larger scale and also higher design complexity comparing to the previous example. Thus, a supportive timber frame, instead of steel local reinforcements is used to stabilize the paper-tube structure.

- Fasteners

Another example of bolted lap joints protected with rubber caps, implemented on bent tubes for outdoor application, is presented as a case study developed by researchers from the field of civil and environmental engineering (Preston, 2011).

A case where threaded steel rods are used to align paper-tubes in a row, to form a wall (fig. 2.27) and nuts are used at the ending points, to apply pressure, appears in the emergency-shelter project referred to as 'Paper log houses' (McQuaid, 2003, p. 34).

Several examples identify use of screws in combination with wood inserts to fix paper tubes together or to other structural elements (floor slab, fundament or roof). A representative application regards the use of paper-tubes as columns, like in the reference project 'Paper house at lake Yamanaka' (JPN) (Miyake, 2009, p. 44), built in 1995. Further details on the performance of the screws in this kind of joints are discussed in case study 4.2.1.1 'Timber plate puzzle node'.

Form-locking

Form-fitting between the tubes creates good conditions for the minimization of extra elements in the assembly. A very basic example appears in a project called 'Paper Partition system 4' (Jacobson, 2014, p. 12). It is a simple free-standing orthogonal frame structure, built indoors. Due to the low structural requirements the form-locking connections could be dry. Only for the linear form-fitting connections, realized with placing small paper-tubes as inserts, tape is applied from the outside, to ensure that they won't become detached because of tensile forces along the length of the tubes. A complete temporary emergency shelter built with form-locking joints is a 'Paper Log House' project (Jacobson, 2014, p. 269). The main building system is shown in figure 2.25. Tubes with different diameters, shaped for form-fitting are glued together and further secured with simple ties. Pieces of paper-tubes are laminated on the outer side of co-linear tubes to secure them together. A case study on this subject developed as part of this research, with focus on the optimization of the form-fitting, is presented in the appendix (6.4.2.1).

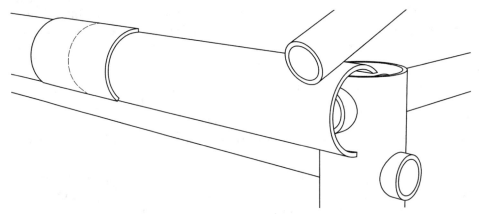

Figure 2.25 Detail of form-locking joint fixed with tie-wraps, drawn based on project: Paper Log House , designer S. Ban (2016, in PH)

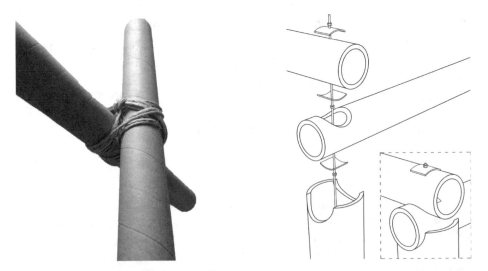

Figure 2.26 Basic joining techniques briefly investigated in this research: tied rope joint – bamboo style node- (left) and form-fitting pressure joint fixed with bolts (right).

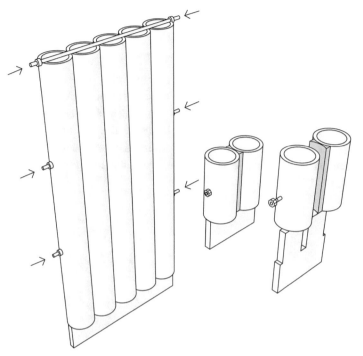

Figure 2.27 Detail of bolted joints used to assemble a wall, drawn based on project: Paper Log House, designer S. Ban (2000, in TU).

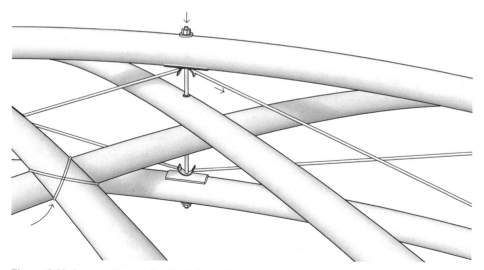

Figure 2.28 Construction detail of bolted steel joint with spacers, tensile cables and tie wraps used occasionally to prevent movements, within a multi-layered grid-shell, drawn based on project: Paper arch, 'The museum of modern art', S. Ban, 2000, US (McQuaid, 2003, p. 68).

Hybrid techniques and intermediate joints

Light timber nodes

The name indicates that such nodes are composed of multiple thin plates assembled together. There are some different possibilities to form these assemblies. As shown in fig. 2.29 an option is to laminate them together -this is the weakest assembly- or to create slits and interlock them. Another possibility is by using dowels, as shown in fig. 2.30 or use metal plates to screw them together. Forming a cross per direction is essential to get sufficient stability. Such examples are presented in temporary arch shelters (Jacobson, 2014, pp. 172, 206).

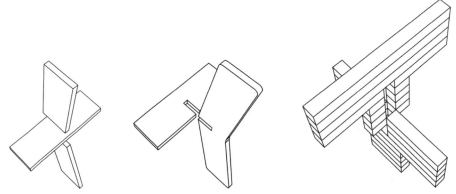

Figure 2.29 Intermediate light wooden joints, an easy solution for basic assemblies. Three basic methods of assembling a light-wooden node are demonstrated. From left to right: gluing, interlocking, laminating slender timber beams & interlocking (Miao Miao paper Nursery school (Jacobson, 2014, p. 265))

There are numerous examples of joints that fall under this category, a few of which is referenced below. Systematically these are part of temporary structures that develop on the level of the ground only.

In the humanitarian project 'Paper Log House, 1995' (Jacobson, 2014, p. 93), with houses of about 16m² surface, the roof is supported by such joints, as shown in the details of fig. 2.31. The same roof design is also implemented in another sheltering project, where the building system for the walls is made of wooden plates interlocking in an orthogonal grid (Jacobson, 2014, p. 187).

The same concept has often been used for foot-joints (fig. 2.37), as in the 'Nomadic Museum' in New York (Miyake, 2009, p. 203) and the project 'Paper Church' (Jacobson, 2014, p. 102). In this case, the cross is fixed with the tubes with screws. Considering the thickness of the plates, on which the screws are applied, the areas close to the fasteners are critical for failing. Some series of structural tests (McQuaid, 2003, p.

77), focused on this aspect, where by applying compression on the paper-tube a shearing effect is created at the area of the joint, showed that the tubes deform significantly at the area of the head of the screw, a fact that leads to vertical displacement of the tube.

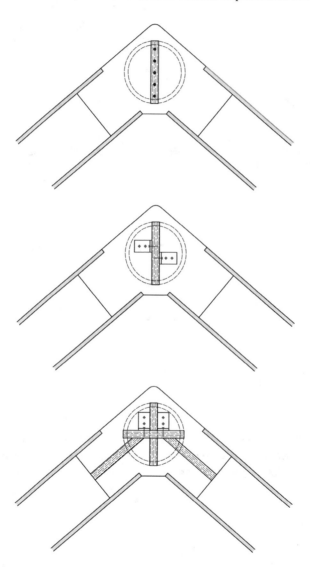

Figure 2.30 Detailed assembly methods for light wooden nodes, combined with dowels and fixed steel plates, sketched in the process of designing the joints for 'House 1' (see fig. 3.4).

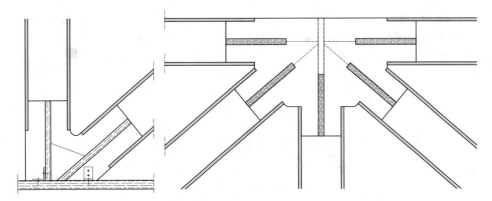

Figure 2.31 Details of roof joints, with diagonal beams integrated, drawn based on the project 'Paper Log House', S. Ban (1995, JPN).

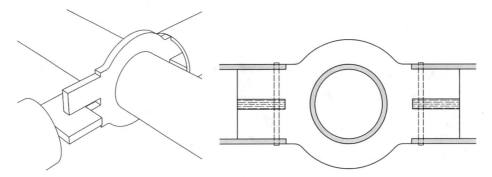

Figure 2.32 Details of joints for the integration of secondary roof-beams, drawn based on the project: 'Chengdu Hualin Elementary School', designed and built by the team of S. Ban.

In a different way, in Chengdu Hualin Elementary School (Miyake, 2009, p. 148), (see fig. 2.32) the concept of an interlocked cross is used to form a ring that surrounds the profile of a main beam, in order to densify the grid of beams only in one direction. This way, the 'ring principle' is used for the secondary stiffening beams that prevent bending of the main ones. Hence, the number of complex joints is minimized. When combining joining principles in such a way, the aspect of tolerances is important, to prevent mis-fitting, in this instance with the roof plates lying on top.

Wooden plates have more functions in paper-tube structures. As the shape of round profiles leads to limited contact with surrounding surfaces, plates with a straight and a curved side have occasionally been used to improve the distribution of forces (Jacobson, 2014, p. 252).

Massive wooden nodes

This category includes joints based on wooden blocks. It creates the possibility for the tubes to transfer forces along the entire perimeter of the profile, due to a massive wooden core that is always in the middle of the node.

Perhaps the most simple approach appears in the project 'Library of a Poet', (McQuaid, 2003, p. 18), in which, as part of an orthogonal grid, single blocks of wood are used to fix steel bars that go through the tubes, as shown in fig. 2.33. This way the joints contribute in prestressing the tubes and keep them under pressure, together with the bracing cables. A similar node made of steel is presented in an outdoors roof structure (Miyake, 2009, p. 31).

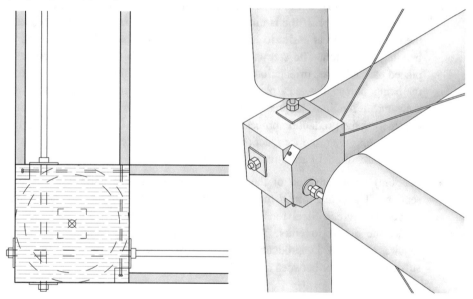

Figure 2.33 Connection details with wooden blocks and integration of steel threads, drawn based on the project: 'Library of a poet', designed by S. Ban (1991, JPN).

A different approach regards plug-in massive wooden joints. In this case, on each side of the central block a cylindrical insert is integrated, on which the tube is fixed, with mechanical fasteners. When the tolerance between the tube and the joint is between 0-1,5mm approximately then friction also prevents movements.

There are several approaches about how to form such nodes, either with natural wood or other timber products. These are defined by the raw material in use, in combination with the assembly methods required to form the node. These two aspects are faced as parameters that lead to different steps in the production process and also a final product with a different matrix.

The most known methods are presented in the construction details in fig. 2.34 - 2.36 and are briefly described hereby:

- (Top) - Blocks of natural wood or timber fitted together with basic interlocks and fixed with bolts. The dimensions of the raw material imply restrictions regarding the diameter of tubes for which its size is sufficient. A reference project is the 'IE Paper pavilion' that is perhaps one of the most successful examples of a paper-tube structure in terms of aesthetics (Shigeru-Ban-Architects, 2013).

- (Middle) - Pre-cut, laminated, multi-layered timber plates. The single plates are easy to cut in shape. If the lamination is not performed in a controlled process, then delamination is easier to happen, if for instance the node is subjected to bending or torsion. The screws that fix the node with the tubes are actually placed parallel to the orientation of the fiber and so occasionally screws can end up being placed between plates (poor anchoring). Thus, the integration of mechanical fixations to press the plates together is important. Furthermore, considering the production process, significant amount of glue is needed for the lamination, when the production of the timber plates already requires a significant amount of adhesives and energy.

- (Bottom) - Combination of the two methods above: the node is divided in sections. These are joined together with biscuits (dowel-type fasteners from timber structures). Steel rods and metal plates are strategically integrated to secure the blocks together. This solution includes many additional cuts that require precision in production and increase the importance of ensuring sufficient distances between all fixation elements.

For the project 'paper dome', where the second type of node applies, as these are described above and is also presented in fig. 2.35, compression and bending structural tests for the joints were performed (McQuaid, 2003, p. 78). The paper tubes used have an outer diameter of 29cm and wall thickness of 20mm approximately. Based on the three-point bending test, the importance of planning sufficient distances between the fasteners is highlighted, as the specimens failed because of cracks appearing on the tubes, right along the lines of screws. Similar testing processes are followed in this research and thus the main outlines for the testing set-up can be seen in paragraph 4.2.

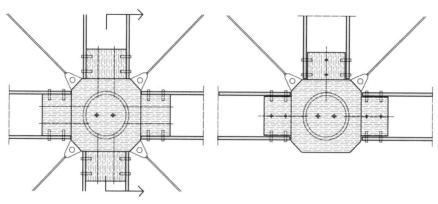

Figure 2.34 Intermediate massive wooden joint fixed in the tubes with bolts, drawn based on the project 'IE Paper pavilion', designed by 'Ban architects'.

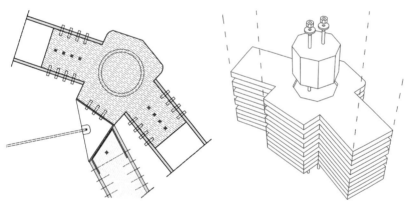

Figure 2.35 Intermediate multi-layered massive timber node, drawn based on the project 'Paper dome' (McQuaid, 2003, p. 48)

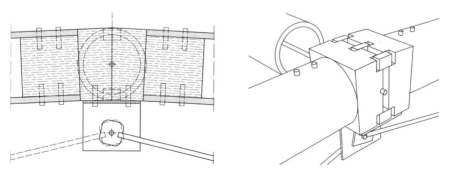

Figure 2.36 Intermediate massive wooden node details, drawn based on the projects: 'Studio' (Paris, FRA) and 'Chengdu Hualin Elementary School', (Jacobson, 2014, pp. 17, 188).

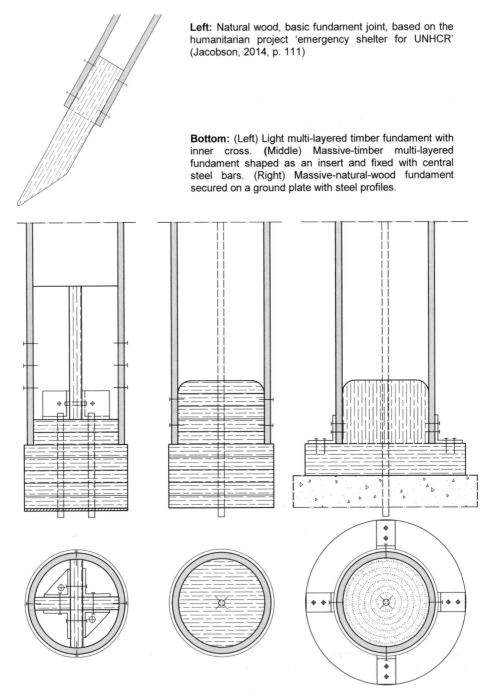

Left: Natural wood, basic fundament joint, based on the humanitarian project 'emergency shelter for UNHCR' (Jacobson, 2014, p. 111)

Bottom: (Left) Light multi-layered timber fundament with inner cross. **(Middle)** Massive-timber multi-layered fundament shaped as an insert and fixed with central steel bars. (Right) Massive-natural-wood fundament secured on a ground plate with steel profiles.

Figure2.37 Common fundament details, based on the review of realized projects.

Steel nodes

For skeletal structures with bigger dimensions than just these of the small shelters that have already been reviewed, steel joints are often implemented. Essentially, the sides of the paper-tubes are capped. Then, a few different options apply regarding the composition of the node. In fig. 2.38 the main assemblies used to pre-stress the tubes and connect them with adjacent elements are shown, based on projects that are presented in the current section. The steel bar is either screwed inside the cap or bolted on the outside. The seam between the tube and the steel cap is sealed. On the cap, either a steel rod is welded (Westborough Cardboard School Building, UK, (Cottrell-Vermeulen-architecture, 2001)), or a steel plate.

Regarding the design of the node, a few different concepts are identified:

(1) Multi-lap hinged node

(2) Flower node

(3) Sleeve node

Further details on their application and assembly method are provided below.

(1) This possibility is used to assemble a pitched roof, as presented in the reference project 'Nomadic Museum', NY, US, (Miyake, 2009, p. 192). To form the multi-axial joint, all steel plates are aligned in parallel planes, in a multi-lap joint and bolted together at the center, as shown in the conceptual detail in fig. 2.41. Joints of the same kind are also part of the roof structure in the project 'Cardboard Cathedral', NZ, (Jacobson, 2014, p. 250).

The number of beams that can be connected with the type of joint described above is in some cases limited. To explain this, when all beams belong initially in the same plane, then the re-arrangement of the beams in slightly different planes, following the concept of the multi-lap joint is not a problem. However, for skeletal structures with more complex geometries this point could lead to misalignments and unnecessary complications.

(2) Thus, for skeletal structures with organic shapes, a different typology of node is identified, occasionally named a s 'flower node', fig. 2.39. The connection is mainly composed of two parts. A central part and all the profiles that are bolted with it. The 'Nomadic paper dome' project, NL, and the 'Vasarely Pavilion', FRA, (Octatube, 2010) make use of this principle. Tensile cables create links between the nodes and enhance the stability of the structural system, together with the steel bars that are integrated in the tubes.

(3) A more complex design is identified in the project 'Boathouse', FRA, (Miyake, 2009, p. 74). A cylindric timber beam is inserted in each tube and screwed.

This assembly is then inserted in a shell made of steel which is the key-element of the node. All elements are screwed together. In the center of the node a flat cap is integrated on both sides (inside and outside) that allows for easy attachment with other structural elements (building-skin and secondary supportive frame). Even though from an aesthetical point of view this node presents advantages, its production requires remarkably higher effort comparing to the previous ones. Moreover, even though the node itself could be stronger comparing to type 2, as addressed above, (more surface for distribution of stresses), it is uncertain if this fact creates any benefits for the paper-tubes, especially as these are not prestressed.

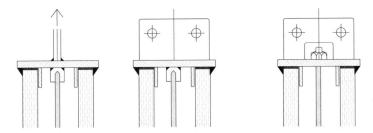

Figure 2.38 Different techniques for the fixation of steel bars through paper-tubes. This detail is reproduced based on the original sketch designed by Octatube engineers (Octatube, 2010). The use of it is in pre-stressed truss-like structures that use round paper-tubes as load-bearing profiles.

Following the previous categorization of nodes in three groups, a selection of pictures from reference projects is presented, to discuss the assembly process. The three projects presented in figures 2.42 - 2.44 show implementation of the 'flower node'. The design principle supports a rather simple manufacturing process, by welding the steel elements together and allowing for adjustments during assembling the bolted joints. In addition to this, as it can be observed in the images, the structures were partly pre-assembled in a few units that were later connected together on the site. As the joints are not fragile and the tubes are pre-stressed the transportation of the units is expected to be effective with limited danger for damaging the construction elements. Next to this, the steel nodes make it easy to attach further elements, such as a protective outer skin.

In general, the idea of implementing solutions that are so highly durable is that the joints could be re-used multiple times whereas the tubes are replaced over time, depending on the specific conditions that apply.

The project 'Ring Pass Delft' (fig. 2.45) presents a different approach, with a more advanced solution, by integrating all elements of the joint in one design-piece. In this case, the design of the joint aims to create an effective multi-plug that can be used for the total of joints within the roof structure and is ideal for truss-structures. Further than that, the aspiration of the aesthetical value in the design is clear.

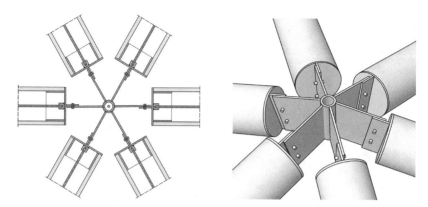

Figure 2.39 Details of steel flower-node, based on the projects 'Nomadic paper dome' and 'Vasarely pavilion', designed and constructed by a group of architects and engineers.

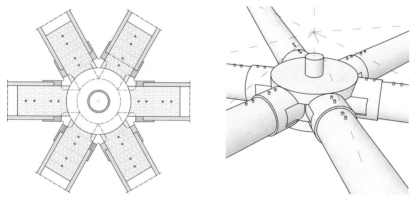

Figure 2.40 Conceptual details of a steel sleeve-node based on the project 'Boathouse', (Ban, 1998, FRA)

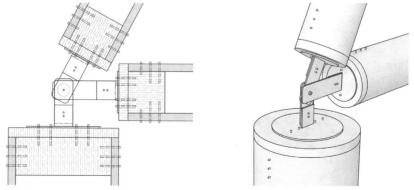

Figure 2.41 Details of a steel node composed of flanges that are hinged at the center. This joint is drawn based on the project: 'Nomadic Museum' (2005, NY, US) by S. Ban.

Figure 2.42 Project: 'Nomadic paper dome' (2004), author: Octatube (NL)

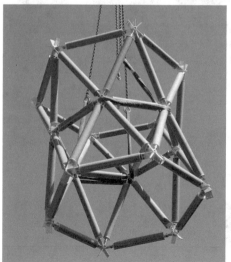 

Figure 2.43 Project: 'Vasarely pavilion' (2006), author: Nils Eekhout (top and bottom right) and Octatube (NL) (bottom left).

Figure 2.44 Project: 'Paper bridge' (2007), architect: Ban S., realization in cooperation with Octatube & other parties, author: Octatube (top) and Theo van Pinksteren (bottom)

Figure 2.45 Project: 'Ring Pass' (Delft, 2010), author: Octatube (NL)

3d printed nodes

Such nodes are preferred when limited structural performance is required and the production of custom 3d-shapes is important to generate the desired geometry. The example presented in fig. 2.46 is based on a realized shelter, an arch-tunnel for 'UNHCR' (McQuaid, 2003, p. 31).

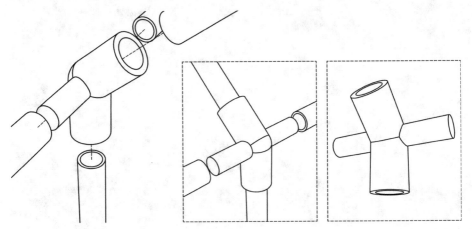

Figure 2.46 3d-printed joint for temporary shelters, based on the humanitarian project 'arch-tunnel for UNHCR' (1999), by S. Ban.

Figure 2.47 3d-printed joint made by N. Kiziltoprak (ISM+D), for educational purposes, in line with the investigations of the joints designed for 'House 1' (see fig. 3.4), as part of this research.

Concrete nodes

Even though traditionally concrete is used in massive construction, it's also popular for casting custom shapes. A basic mold and a suitable mixture of aggregates (this will define the material properties) are sufficient for the production and steel reinforcement is also easily applicable. Such an example is presented in the 'Paper Pavilion' (SG), Biennale 2006 (Miyake, 2009, p. 102).

Of course, even in paper-structures concrete is one of the standard options for the fundament. Optimally, the paper-tubes are kept at a distance from the ground (Miyake, 2009, p. 31).

2.2.4.2 Assembling L, U & square paperboard profiles

Figure 2.48 A collection of flat, L and U profiles that are readily available in the market.

Based on the literature review, so far, none of these products has been used in any construction project with specific safety requirements. At the same time, the level of information on the aspect of structural performance of the profiles or assemblies is poor. They are mostly used in the packaging sector, to reinforce boxes, function as spacers or protect parts of the content. They are also used to build prototypes of structures and small-scale pavilions. Within the context of this research, student projects in the field of architecture that have as main subject the design of emergency shelters consider the use of these profiles as part of conceptual building systems (Latka, 2017, pp. 376, 385) and (Latka, 2016). A more recent experimental study that discusses L-profile as a beam element, presents an interesting approach about how to build a 'Zollingerdach' (arch-type structure) with bolted fixations reinforced with steel plates (Hartmann, 2019).

The main joining techniques that apply for this category of materials are already identified in table 2.4. Therefore, hereby mostly novel details that focus on potential applications in construction are presented.

L and U profiles are produced by bending flat profiles. The angles formed in these profiles often present significant deviations from the desired 90° angle. Such imperfections can lead to problems in case of lamination of the components and cause delamination.

Fig. 2.49 (left) shows a conceptual detail of a bolted T-profile that can be used to attach lightweight cladding elements on top. A study on the potential of laminated facade panels presents interesting grounds for this application concept (Krüger, 1999). In a different way, in fig. 2.49 (right), the L-profile is used as a protection edge for a panel. In a similar way, a U-profile could be used, as shown in fig. 2.50, to enclose the 'open edges' of a multi-layered paper-based panel.

The futuristic 2d-plan detail of fig. 2.51, shows a concept for a paper-based building system that combines the details discussed above. Spirally winded square profiles with dimensions of 10 x 10 cm section (for instance) and a thickness of 10mm are already available in the market. These present obvious advantages, comparing to the round tubes, when it comes to designing a skeletal frame on which cladding can easily be attached. However, at the same time, the structural performance of these profiles is not yet comparable to this of the paper-tubes.

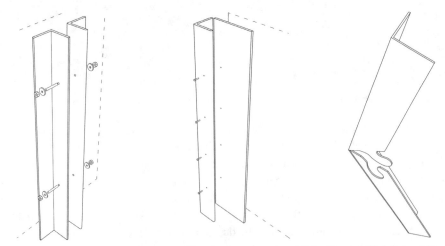

Figure 2.49 L profiles in assemblies. (Left) Bolted assembly of T-Profile, (middle) assembly of a protective edge (pinned), (right) corner interlock - based on edge-protection product.

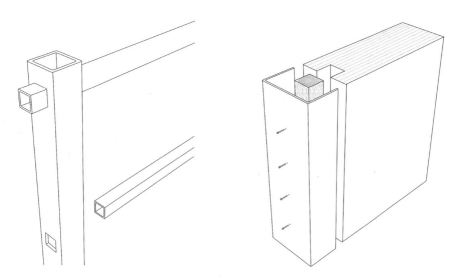

Figure 2.50 (Left) Assembly of square profiles in a frame, (right) finishing of a paper-based wall element with a U-profile.

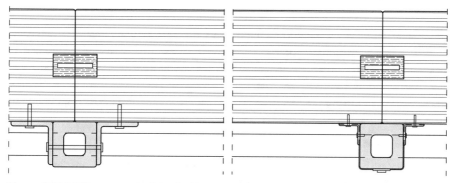

Figure 2.51 2d Plan conceptual detail of façade assembly.

Figure 2.52 Example of haptic prototype for a fully paper-based building system (interdisciplinary educational module on façade technology, winter semester 2017-2018).

2.2.5 Further types of specialized joining techniques

These mainly regard components from table 2.3.3, such as casted and 3d-printed objects. A relevant application, in the spirit of recycling paper materials, is to form light bricks or panels that could be used as insulation or cladding (Diarte). For components with high thickness form-fitting in a 'Lego-brick' concept is a possibility. For components with low thickness that need to be fixed on a vertical surface, fasteners might be the solution, as performed in a research project that investigates the development of façade elements from waste materials (Böttger, 2019). However, for the moment these kinds of paper-based products still need to be optimized before the joining techniques are elaborately developed. The high fiber content in combination with other factors, such as climate conditions, lead to significant changes in the volume and hardness.

2.2.6 Summary

The joining techniques for paper-based products are reviewed by material category. There are no rules or formal criteria about the applicability of these techniques in construction. Similarly, there are no guidelines for the design of the joints. On these grounds, the existing solutions are listed per category (form-locking, force-closure, material-closure), following the literature review, and the main influences from common products and construction practices are identified.

To summarize these:

In the category of basic joining techniques, commonly used adhesives are listed, with PVA being a solution approved for construction purposes. Next to this, similarities with textile assemblies are identified, such as stitching and fastening with snap joints.

In the category of boards, form-locking methods, like in timber structures are commonly used, despite limitations in the structural performance that are mostly related to the occasional matrix of the board and the behavior of the basis material itself. Special screws and dowels for extra-lightweight boards belong to the most promising mechanical fixations.

In the category of profiles, the majority of findings regard round profiles (paper-tubes). A wide range of applications and joining techniques are reported. Influences from bamboo structures are obvious in examples of form-locking connections, as well as from timber and steel structures in examples of intermediate connections.

Based on the review of building methods within realized projects (as explained in 2.1.1), paper-based products (2.1.2) and joining techniques (2.2), the round paper-based profile is, up until now, the most preferred component for experimental load bearing structures. In other words, it is the main paper-based component pictured in the field of construction. The reviewed information on assemblies and joints create favorable conditions for further research. At the same time, numerous issues need to be clarified, as besides the occasional projects and the information referenced, data on requirements and performances are lacking.

Literature

Ayan, Ö. (2009). *Cardboard in architectural technology and structural engineering: A conceptual approach to cardboard buildings in architecture.* Retrieved from

Böttger, W. (2019). *Development of (interactive) facade elements on base of waste materials of (waste) water companies.* Paper presented at the 3rd International Conference on Bio-Based Building Materials (ICBBM), Belfast, UK.

CelluTex. An office desk made of paper. Retrieved from www.cellutex.eu

Conde, I. (2012). Analysis of adhesive joints in corrugated board under shear loading. *Adehsion and adhesives, Elsevier.* doi:0143-7496/$

Cottrell-Vermeulen-architecture. (2001). Westborough Cardboard School Building. Retrieved from http://www.jdg-architectes.com/projet/pavillon-vasarely/?lang=en

Diarte, J. Urban Recycling Intervention: Prototype Tooling for Transforming. Waste Corrugated Cardboard into Architectural Elements and Building Components.

ETSAP. (2015). *Pulp and Paper Industry.* Retrieved from

Fadiji, T. (2016). Compression strength of ventilated corrugated paperboard packages: Numerical modelling, experimental validation and effects of vent geometric design. *Elsevier, 152*, 231-247.

Fiction-Factory. Wikkelhouse. Retrieved from https://bouwexpo-tinyhousing.almere.nl/fileadmin/user_upload/Wikkelhouse_BouwEXPO.pdf

Hahn, E. K. (1992). Edge-compression fixture for buckling studies of corrugated board panels. *Experimental Mechanics.*

Halonen, H. (2012). *Structural changes during cellulose composite processing.* Retrieved from

Hartmann, L. (2019). *Zollinger roof made of paper materials: design, dimensioning and implementation.* Retrieved from

Iggesund-paperboard. *Reference Manual. The Paperboard Product.*

Jacobson, H. Z. (2014). *Humanitarian Architecture.* Aspen, Colorado 81611: D.A.P. (Distributed Art Publishers, Inc.).

Koubaa, A. (1995). Measure of internal bond strength of paper/board. *Tappi, 78*(3), 103.

Krüger, G. (1999). AN ELASTIC ADHESION SYSTEM FOR STRUCTURAL BONDING OF FACADE PANELS. *Otto-Graf-Journal, Vol. 10.*

Latka, J. (2016). House of cards. Retrieved from www.archi-tektura.eu

Latka, J. (2017). *Paper in architecture. Research by design, engineering and prototyping.* (2212-3202). Retrieved from Architecture and the built environment, TU Delft: https://cbonline.boekhuis.nl

Lawrence C Bank, T. D. G. (2015). *Paperboard tubes in architecture and structural engineering: A review.*

Li, Y. (2016). Experimental and numerical study of paperboard interface properties. *Society for Experimental Mechanics, 56*, 1477-1488. doi:10.1007/s11340-016-0184-8

Li, Y. (2017). Anisotropic elasic-plastic deformation of paper: out-of-plane model. *International journal of Solids and Structures.* doi:10.1016/j.ijsolstr.2017.10.003

Linvill, E. (2014). The combined effects of moisture and temperature on the mechanical response of paper *Experimental Mechanics, 54*, 1329-1341. doi:10.1007/s11340-014-9898-7

McQuaid, M. (2003). *Shigeru Ban.* London N1 9PA: Phaidon Press Limited, Regent's Wharf All Saint's Street.

Miyake, R. (2009). *Shigeru Ban Paper in Architecture.* US: Rizzoli International Publications.

Octatube. (2010). Cardboard structures. Retrieved from https://www.slideshare.net/booosting/booosting-27mei2010-octatubecardboard2

Perez, H. (2013). Analysis of twist stiffnee of signle and double-wall corrugated boards. *Composite structures, Elsevier, 110.*

Preston, S. J. (2011). Portals to an Architecture: Design of a temporary structure with paper tube arches. *Construction and Building Materials, Elsevier, 30.* doi:10.1016/j.conbuildmat.2011.12.019

Schönwälder, J. Cardboard as building material. Retrieved from https://docplayer.net/28996964-Cardboard-as-building-material-dipl-ing-julia-schonwalder.html

Schönwälder, J. (2008). *Mechanical behavior of cardboard in constructionCardboard in architecture* (Vol. 7, pp. 131-146). doi:10.3233/978-1-58603-820-5-131

Schönwälder, J. (2016). *Computational modelling of cardboard: an anisotropic constitutive model incorporating creep and rate dependence.* Retrieved from Fracture of nano and Engineeering materials and structures: https://link.springer.com/chapter/10.1007%2F1-4020-4972-2_235

Schütz, S. (2017). *Von der Faser zum Haus*: bauhaus ifex research series 1.

SETIS. Strategic energy technologies information system. Retrieved from https://setis.ec.europa.eu/technologies/energy-intensive-industries/energy-efficiency-and-co2-reduction-in-the-pulp-paper-industry/info

Shigeru-Ban-Architects. (2013). IE Paper Pavilion. Retrieved from https://www.archdaily.com/354471/ie-paper-pavilion-shigeru-ban-architects

TU-Delft. (2008). *Cardboard in Architecture* (Vol. 7): IOS PRESS.

VDI, G. E. K. V. (2004). *Methodical selection of solid connections. Systematic, design catalogues, assistances for work.* Retrieved from

2.3 A review of joining techniques for common construction materials

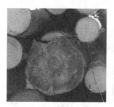

Figure 2.53 Materials reviewed in this chapter with focus on the joining techniques (from left to right): timber, bamboo, textiles and composites, steel.

Considering the global idea of using paper-based components to develop highly recyclable temporary structures, it is useful to think in which ways these components are compatible with common building practices and assemblies. In this section, common joining techniques with implementation in conventional building methods are reviewed, with the aim to expand the spectrum of available solutions with potential for assemblies composed of paper-based components. This review aims to function as a base for discussion, to examine which joining techniques could be effective and also indicate when specific reasons show incompatibility. From a different perspective, when examining the implementation of paper-based materials in construction, it is useful to see if and how known assemblies could be realized also with such components. In this spirit, it is useful to discuss the potential integration of such components in assemblies that engineers are familiar with.

In this context, this chapter presents the outcomes of the review that show how paper-based components present similarities with common building materials, especially natural fibrous materials, textiles and multi-layered composites. Next to this, an opportunity is created to evaluate what is important when joining paper-based components. The most important aspects, within the review process, are in principle related to the impact of the respective joining techniques on the material and the relation between the structure of the material and the effectiveness of the joint for transferring loads and also being durable. The review consists of four main sections, defined by material. Priority is given to timber, as the most directly related material to paper-based products and then bamboo, textiles and composites and steel follow. Each section starts with a short introduction about the material and a brief overview of joining principles which are later discussed under the three main categories of material-closure, force-closure and form-locking. In paragraph 2.4 the outcomes from the entire review process are assembled in an overview and the most important observations are summarized per case.

2.3.1 Timber

Figure 2.54 Natural wood (Black-forest, DE). The material structure is heterogenous, cellular and anisotropic. A cell consists mainly of cellulose, hemi-cellulose and lignin (Wikipedia). The rings visible on the section of the core show the age of the tree.

About the material

There is a close relation between timber and paper products, mainly based on the common origins of both materials and the fibrous material structure as a result. Hence, common joining techniques that apply for timber structures are hereby discussed, focusing on the state of the art in relation with the grain direction, as the fiber orientation is crucial for the design of the joints. For example, mechanical fixations are always applied perpendicularly to the grain direction, whereas in form-locking joints the opposite happens. In table 2.7 basic joining techniques are presented that are discussed in the review.

Traditionally timber is considered a highly durable construction material, due to its structural performance under loading and also fine preservation through alternating climate conditions. It is most commonly preferred for low-rise buildings or integrated in hybrid structures, so in combination with either bituminous materials or steel. More specifically, it is used for the production of a wide range of construction elements, such as beams, profiles for openings, boards for flooring and cladding etc. Additionally, it is a very popular material for interior design. Based on its natural and renewable wood origins, timber products are established as ecologically friendly. On the other hand, some downsides are identified in the significant energy required to properly engineer the wood, the 'end of life' process that most of the times leads to downcycling and also the intensive use of adhesives.

Regarding the mechanical behavior, timber is an anisotropic and more precisely ortho-tropic material. Its strongest dimension is parallel to the orientation of the grains. Hence the mechanical properties of timber products are characterized in all three directions, but sometimes, in engineering, simplified material models are considered for the analysis of draft structural models. In general, properties differ significantly depending on the raw material and also the manufacturing process.

Figure 2.55 Timber is used for many different applications in construction, both as a load-bearing and cladding material. Timber products are available in a great range of qualities and colors. (Left) dome structure built with natural timber and steel joints (TU Delft, Botanical garden), (right) timber façade-cladding (Delft, residential/ office building, train-station area).

Joining techniques

Material-closure

Adhesives are widely used within the production of timber products, such as plywood, particle boards or cross-laminated timber-plates for construction (fig. 2.56). In the case of laminated plates and beams, the seam between the subsequent pieces is commonly a finger joint (table 2.7, 2b). This joining method is often used to produce complete sections for timber houses, bridging this way the eave detail, as shown in examples of timber construction manuals (Herzog, 2003, p. 120), (Borgström, 2016, p. 106).

Overall, there is a wide variety of adhesive products that are suitable for certain wood products and can be applied in specific conditions (temperature). An informative over-view of 'adhesives in the wood industry' is provided by Manfred Dunky, where the main types of adhesives are figured in page 4, table 1 of the publication (Dunky, 2017). Re-garding the ecological aspect, a study on 'green binders for wood adhesives' analyses the potential of bio-based possibilities and addresses the issues that slow down the establishment of green solutions such as water resistance (Emelie Norström, 2018).

Table 2.7 Basic joining techniques for timber.

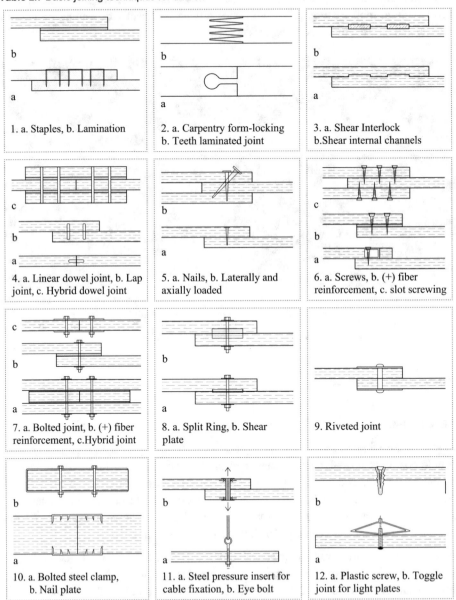

1. a. Staples, b. Lamination	2. a. Carpentry form-locking b. Teeth laminated joint	3. a. Shear Interlock b.Shear internal channels
4. a. Linear dowel joint, b. Lap joint, c. Hybrid dowel joint	5. a. Nails, b. Laterally and axially loaded	6. a. Screws, b. (+) fiber reinforcement, c. slot screwing
7. a. Bolted joint, b. (+) fiber reinforcement, c.Hybrid joint	8. a. Split Ring, b. Shear plate	9. Riveted joint
10. a. Bolted steel clamp, b. Nail plate	11. a. Steel pressure insert for cable fixation, b. Eye bolt	12. a. Plastic screw, b. Toggle joint for light plates

There are no universal standards yet for adhesive bonds in timber structures. Each product comes together with technical specifications and testing results for a number of assemblies are accessible. The specific products used for experiments within this research are mainly addressed in paragraph 4.2.

Lap joints are occasionally reinforced with an adhesive bond, sometimes with integration of fibers in the mixture, to increase the strength of the bond directionally.

Figure 2.56 Cross laminated timber (CLT) construction, in the building process, at the area of Delft train station (NL). The prefabricated CLT panels are installed on the site with steel plates and numerous fasteners. The structure is assembled on a continuous concrete fundament. This simple design could be a great base for discussion for researchers considering multi-layered paper-based panels for construction purposes.

Figure 2.57 (Left) Implementation of CLT on multi-storey buildings with balconies (area of Delft train station (NL)). (Right) Extensions that exceed the perimeter of the main building are supported by laminated timber-columns, mounted on steel-feet and separate concrete foundations.

Force-closure

Mechanical fixations encounter the most common joining techniques for timber construction. An overview of the most common basic principles is shown in table 2.7. Ground rules for essential dimensioning and design according to the fiber orientation are described by standards and more specifically, in EN 1995-1-1, section 8 (Leijten). The main categories of connectors as described there are the following:

- dowel type fasteners
- threaded bolts
- punched metal plates
- shear plates
- split rings

Next to this, further guidelines with specific conditions may apply for different countries, especially in countries that produce timber products that provide further design values and construction details (American-Wood-Council, 2008). Similarly, a wide variety of guides and manuals is available for form-locking joints, also known as carpenter joints. Another such example is cross laminated timber, a relatively modern material that is not included in Eurocode 5, but specifics (including the aspect of joints) are defined in more recent manuals (Harte, 2017).

Moreover, extensive research to identify the load-capacity of fasteners in different types of wood and for different failure modes is constantly developed, with the aim to clarify conflicting opinions and develop optimized models for the evaluation. Due to the fibrous non-homogeneous material structure and also the wide variety of products the prediction of the structural behavior of joints is still a process in development, as relevant research indicates. The main reason, indicated in studies about multi-fastener joints, is that often the failure is demonstrated on the timber (Pouyan Zarnani, 2014). It is expected that similar problems would rise when testing similar joints on paper-based assemblies. Thus, experiences from wood can be very helpful to define methods for predicting the structural behavior and also perform more effective structural tests.

Overall, the great advantage of this group of connections is that the design of the joints can be effectively adjusted according to the special needs, meaning the forces that need to be transferred. Therefore, there are numerous examples of different designs for joints that transfer compression, tension, shear or moments. Below the main types are discussed and their potential application on paper-based products is considered.

- Dowel type fasteners

These connectors are used to form direct joints between timber elements.

In practice nails (table 2.7, 5) and staples (table 2.7, 1a) are the most simple and fastest assembled joining elements. Staples are mainly used to connect boards (with limited thickness) and have the advantage that they can be placed very close to each other (K.C. Talbot, 2008). Nails can handle forces applied from the sides that create shear effects, but are problematic when subjected to vertical forces that cause pressure or tension. Nails fixed in an angle (slanted) can improve this behavior but complicate possible desired destruction of the elements.

On the other hand, screws present the opposite behavior, as they have better grip that leads to higher withdrawal capacity, but are more sensitive to side forces. Next to this, a wide spectrum of designs for screws is provided, with differences in the areas of the spiral and the head that are suitable for different cases, providing optimized performances against the failure modes as these are described in EN-1-1: Section 8. A special category of screws that is interesting for small scale structures is slot screwing, a reversible joining method (VET-WA, 2013, p. 23).

Wooden dowels are placed internally and are effective for pressure joints. Similarly, biscuits (inserts of elliptical shape with wider surface) are often used to connect boards.

Bolts solve the problem of withdrawal pressure (figure 2.58). However, there is a limit in the torque that can be applied, depending on the type of timber. This point is particularly interesting for the application of bolted joints on paper products.

A downside of metal dowels is that they can cause splitting of the material around the connector, as a result of forces applied perpendicularly to the grain that can consequently lead to failure of the joint. Therefore, further techniques are often applied to improve this issue, as described below.

Figure 2.58 Eave details (shelter in Black Forest, DE), a. (Top) Bolted double lap joints in a symmetrically designed structure that prevent bending, (bottom) hidden dowels, a type of joint that functions well under pressure

- Metal plates

Metal plates are used to reinforce the areas of the beams and especially the connections (butt joints).

Perforated metal plates are attached with fasteners and are particularly useful in large span timber structures (Crocetti, 2016, pp. 9, 11). They provide great support, improving the structural performance especially against tensile forces and are therefore widely used to form in-direct joints within structural frames. Commonly failure appears around the perimeter of the connectors' grid, following the same critical issues as described in the category of dowels. Potential implementation on paper-based structures is considered possible.

Punched or nail metal plates are a more basic form of the same principle connection method (Borgström, 2016, p. 139). They are directly hammered on timber, a process that creates significant impact on the material itself. Thus, this joining method would possibly cause damage on paper-based components and therefore is not recommended. Next to this, the implementation of this type of connector is only possible on components with density up to 500 kg/m^3 (Hover 2017).

An example of an indirect steel joint for a hybrid timber structure appears in figure 2.59, whereas a representative example of a metal connection with slotted plates is shown in figure 2.60.

- Shear plates and split-rings

Both types of connectors are used in lap joints, placed internally, between two components. Due to the high precision required to form the grooves in which the connectors are placed, none of these methods is recommended for connecting paper-based components. As paper-based products are in principles much softer than timber products it is expected that such fittings would tear the material and fail.

Overall direct mechanical fixations are often simpler to realize and less costly. On the other hand, indirect reinforced metal joints are often stronger and more durable.

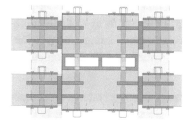 

Figure 2.59 An example of a hybrid timber beam structure reinforced with steel profiles and plates (SWAT studio, TU Delft, 2015-2016, Q1). The steel plates are fixed on the timber with inserts and bolts. Implementation of inserts in paper-based beams would rise questions about causing damages on the component depending on the lamination-plane or while creating the cut-outs.

Figure 2.60 Bolted steel nodes, Louis Vuitton Foundation (Paris, FRA). The nodes are integrated on the beams with cuttings created close to the ends, parallel to the grain direction, filled with the steel inserts. Combination with pressure plates, on the outer sides, would minimize the stresses occurring on the timber. Such a concept could also be considered for multi-layered paper-based beams.

Form-locking

These methods are popular in small scale structures (wiki-house), Japanese joinery, interior design and furniture. A popular example of this kind is the 'Tamedia office' (Shigeru-Ban, 2013).

Traditionally the production requires advanced prototyping skills. Most experiences are known from wood-working manuals (Fearn, 2013). In principle they are easier to make with soft-wood. Modern technologies and specifically CNC Milling allow for fast and precise production, so that even more complex parametric designs are possible (A. Menges, 2019).

Traditionally joints of this group don't make use of any metal elements. All timber parts are embedded together, forming a stable connection that relies on form-locked geometries and friction. However, numerous customized solutions include integration of adhesives or even fasteners to form more secure connections. Glued-in rods are sometimes used as reinforcement.

Simple dry form-locking lap joints are widely applicable on flooring systems and sometimes on wall panels, or stacked beams (Drexler, 2014). Hence, form-locking is often used as a complementary joining method, to create a relatively stable surface for the implementation of mechanical fixations. It could function in a similar way within paper-based structures.

More complex designs appear in frame structures, in the areas of the nodes. The intermediate wooden joints as presented for paper structures in paragraph 2.2.4.1, fig. 2.34 - 2.36, borrow the principles from these joining methods. The performance of different types of form-fitting joints is still a subject in development. The performance under bending is considered to be of the most critical aspects (Kohara, 2004).

Figure 2.61 'Japanese pavilion', installed at the faculty of architecture during the summer semester of 2017 (TU Delft, NL). The assembly is realized with dry form-locking joints. An incredible number of joints is incorporated in the design to achieve effective distribution of loads and consequently stability. It seems that tolerances are integrated in the design of the joints (for example around the corners of the holes on the wall-plates). This way, the pavilion could potentially be disassembled and relocated, keeping in mind that it is designed for indoor-use. A similar design concept would be less functional with paper-based components, as these would become softer over time.

Figure 2.62 Implementation of Japanese joinery details on 'the Euro-palette pavilion' team-project, designed during the Bucky Lab studio (TU Delft, 2014-2015, Q1-2). (Left) Hand-crafted prototype using revived timber, (right) CNC Milled MDF plate. Form-locking joints require advanced skills in prototyping. In the case of paper-based components form-locking joints that are subjected to moderate pressure only, could be functional. However, precision in fabrication would be crucial.

Summary

All in all, joining techniques from timber structures offer a variety of methods and knowledge that can be linked to future developments within research on paper-based structures.

To explain the reasoning while considering the suitability of different techniques, it is helpful to analyze the material structure. As in timber construction the grain-direction is important for the design of the joints and the implementation of connectors, so is the matrix of the paper-product. For example, in the case of multi-layered paperboard, two scenarios could be addressed. Either all the boards are laminated with the primary-fiber-direction (as defined by the forming process) aligned or they are cross laminated. The finished multi-layered boards present, in the first case, similarities with a natural wood board and in the second case with cross-laminated timber. In this spirit, the joining techniques that apply for these two different kinds of engineered wood-products could be respectively considered for the multi-layered paperboards. Keeping this point in mind, the discussion over the suitability of specific joining techniques follows.

On the subject of adhesive bonding, the referenced sources provide information that is useful to consult, both in relation with the available products and also the ecological aspect. The second one raises great concern, due to the fact that when adhesives are involved in the production of construction then huge amounts are needed, affecting significantly the environmental impact of the end product.

Regarding the category of mechanical fixations, the following joining techniques are recommended:

The use of stapled joints is encouraged for simple lap joints and 90° corner joints for which a connector with depth of 14 mm is sufficient.

Nails are recommended when the required depth of the connector exceeds the maximum offered by staples. The reason for this is that staples anchor better in paper-products. However, nails present higher strength against shearing.

In principle, screws with wide spiral and head are more likely to anchor better in paper-based products and thus shall be preferred over thinner designs.

For dowels, wooden dowels are recommended instead of metal ones, to prevent tearing of the paper-based components.

Bolts with pressure plates are recommended as well, always with the condition that testing of custom assemblies is required. Local reinforcement of the components at the area of the joint might be necessary. Perforated metal plates could be applied, whereas considerations are the same as for the use of bolts.

Similarly, biscuits with wide surface are recommended as well, to align elements within an assembly. The combination with glue might be necessary if the aim is a stable connection. As this is an internal connector, normally there is no water access to it. Thus, the selection of water-soluble glue can be considered, if it satisfies the structural requirements.

Direct dry form-locking joints could function for components that are only subjected to moderate compression forces (for example flooring systems). In other cases, combination either with adhesives or fasteners would be necessary (for example supported wall elements). The main concern here is that the combination of 0 tolerances, as required to form a joint of this kind, with the pressure forces in the form-fitting area, would not lead to a long-lasting solution, as softening of the components is expected to occur with time.

Within the review, literature sources provided useful insights about different building systems and also representative construction details for the areas of the fundament and the eave that are good sources of inspiration when considering the design of paper-based structures. Some of these are namely: (Herzog, 2003), (Borgström, 2016, chapter 4) and (American-Wood-Council, 2008, chapters 10 and 11).

Literature

A. Menges, J. K. (2019). BUGA Wood Pavilion, ICD Research Buildings / Prototypes, Bundesgartenschau Heilbronn, Germany. Retrieved from https://www.icd.uni-stuttgart.de/projects/buga-wood-pavilion-2019/

American-Wood-Council. (2008). *ASD/LRFD Manual for Engineered Wood Construction.* 1111 Nineteenth St., NW, Suite 800, Washington, DC 20036, US: American Forest & Paper Association.

Borgström, E. (2016). *Design of timber structures. Structural aspects of timber construction.* SE 102 04 Stockholm: Swedish Forest Industries Federation, Swedish Wood.

Crocetti, R. (2016). *Large-Span Timber Structures.* Paper presented at the Proceedings of the World Congress on Civil, Structural, and Environmental Engineering (CSEE' 16), Prague, Czech Republic.

Drexler, H. (2014). *Timber Prototype.* Retrieved from Leonardo Campus 5, 48149 Münster:

Dunky, M. (2017). *Adhesives in the Wood Industry.* Dynea Austria GmbH, Krems, Austria: Taylor & Francis Group, LLC.

Emelie Norström, D. D., Linda Fogelström, Farideh Khabbaz, Eva Malmströ. (2018). *Green Binders for Wood Adhesives (chapter 4)*: IntechOpen.

Fearn, C. (2013). *The City & Guilds Textbook: Level 1 Diploma in Carpentry & Joinery* City & Guilds.

Harte, A. M. (2017). Mass timber - the emergence of a modern construction material. *Journal of Structural Integrity and Maintenance, 2:3,* 121-132. doi:10.1080/24705314.2017.1354156

Herzog, T. (2003). *Holzbau Atlas.* Basel, Boston, Berlin: Birkhäuser Edition Detail.

K.C. Talbot, L. D. R., C.P. Pantelides. (2008). *Structural performance of stapled wood shear walls under dynamic loads.* Paper presented at the The 14th World Conference on Earthquake Engineering, Beijing, China.

Kohara, K. (2004). A study on experimental testing of joints on timber structures. *13th World Conference on Earthquake Engineering, Vancouver, B.C., Canada*(Paper No. 2441).

Leijten, A. EN1995-1-1: Section 8 - Connections. Retrieved from https://eurocodes.jrc.ec.europa.eu/doc/WS2008/EN1995_5_Leijten.pdf

Pouyan Zarnani, P. Q. (2014). Wood Load-Carrying Capacity of Timber Connections: An Extended Application for Nails and Screws. *Materials and Joints in Timber Structures, Springer,* 167-179. doi:10.1007/978-94-007-7811-5_16

Shigeru-Ban, A. (2013). Tamedia Office Building. Retrieved from https://www.archdaily.com/478633/tamedia-office-building-shigeru-ban-architects

VET-WA. (2013). *Join Solid Timber, Learner's guide, BC2013*: Department of Training and Workforce Development 2016, Government of western Australia.

Wikipedia. Wood-structure. Retrieved from https://en.wikipedia.org/wiki/Wood#Water_content

2.3.2 Bamboo

Figure 2.63 Bamboo, the material.

About the material

Bamboo is a very old-fashioned structural material that grows mostly in tropical climates, in parts of Latin America, Africa and Asia. It is an anisotropic material with high tensile strength due to the fiber orientation parallel to the length-axis, whereas its compressive strength depends highly on the diameter. Next to this, buckling is identified as a serious problem (Raj, 2014). Considerations regarding its implementation in construction are often related to significant shrinkage as a result of water loss. Naturally the durability of bamboo is relatively low and more specifically up to five years, if not treated (Natural-Homes). Further than the similarities between bamboo and paper on a material-level, bamboo structures are a very inspiring source of information for paper-tube structures, a subject that is in the main focus in the second section of this research (see paragraph 3.1.2). Building with bamboo is traditionally a craftmanship. The light weight and easiness in processing create friendly conditions for skillful makers. Hence, essential traditional forming methods and tools about how to shape the joints are available (Lopez, 1981a). Table 2.8 presents some concepts of joining techniques, as a summary of the findings from the review of literature and reference projects. A selection of elaborated details follows.

Table 2.8 Overview of joining techniques for bamboo structures.

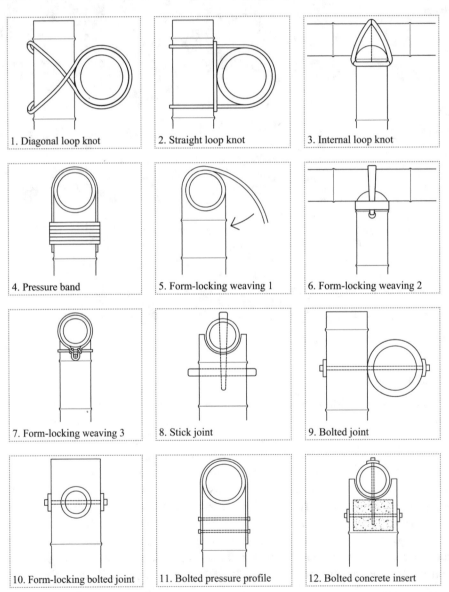

1. Diagonal loop knot	2. Straight loop knot	3. Internal loop knot
4. Pressure band	5. Form-locking weaving 1	6. Form-locking weaving 2
7. Form-locking weaving 3	8. Stick joint	9. Bolted joint
10. Form-locking bolted joint	11. Bolted pressure profile	12. Bolted concrete insert

In general, building with bamboo has been a subject developed within hands on, practical approach that evolved mostly by experience rather than elaborate engineering methods. Hence, there are no standards for bamboo structures, like for timber. Its rapid pace of growing has nowadays brought it once again in the foreground, as researchers try to reinvent its application in construction. Traditionally it is used to build skeletal structures, roofs, partition walls and furniture (Gumangan). In modern times it has been used for semi-indoor spaces, whereas examples of pedestrian bridges in Colombia also show its potential (Kaminski, 2016). Currently the production of timber-like boards is also investigated, especially for countries such as China where timber is lacking (Xiao, 2019). Some modern examples of application include the pavilions for an exhibition in Shanghai (Heinsdorff, 2010) and Mexico (Cogley, 2019).

Joining techniques

Material Closure

Adhesive joints implemented on a form-locking based connection are some very special examples of customized glued joints that use the structure of bamboo material at its maximum. Using the technique shown in page 17 of the guide for constructing with bamboo by Lopez O. the bamboo stick is split into fibers close to the edge. These fibers are then weaved around the form-locked connection. Traditionally the woven connection would be dry, but in modern designs adhesives are used to form a rigid fixed joint. Additionally, other fibers are also often used to reinforce the adhesive bond. Moreover, to simplify this process, often the bamboo pipes are just shaped properly for the form-fitting and textiles are laminated to form-the joint. Most times these are natural fiber textiles. Fiber glass is also an option, but less popular for this application due to its ecological downsides. The binder is often a type of resin. The recyclability of this method is one of the drawbacks. A representative application for this joining technique is to form bamboo frames for bikes. Relevant studies developed in this research are presented in paragraphs 4.2.2.2 and 6.4.1.3 (appendix).

Other reported methods of adhesive joints come from application of bamboo sticks as underground water pipes, where the sealing of the joint is crucial. The main adhesive used there is tar mixed with cotton, whereas three main assemblies are addressed, as shown in figure 2.64. The form-locking between the pipes is essential. Then the tolerance is filled with the adhesive. Joints without form-locking include a metal or plastic pipe instead, placed either internally or externally.

When tar is exposed to sunlight, its temperature can rise significantly and can affect materials in contact. Thus, this mixture could not be used in combination with paper-products. Next to this, tar is never applied indoors.

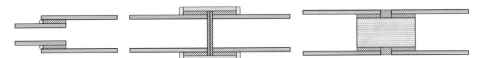

Figure 2.64 Details of linear joining techniques, based on examples from underground bamboo pipes as presented in a guide for building with bamboo (Natural-Homes, p. 48).

Force closure

The most traditional technique used to form a stiff structure with bamboo sticks is to tighten them together. There are numerous tightening methods and also knots, to finish the loop. In the overview provided in table 2.8 some of the most popular concepts are presented. The ropes are commonly weaved around the perimeter of the connection, sometimes using shortcuts that go through the core, or even wooden dowels or pieces of bamboo sticks to secure the elements together and also fixate the ropes (table 2.8 examples 3, 7).

An alternative is to use textiles instead of ropes and tension them around the connection between beams that are fitted together with form-locking. This method is particularly effective for corner eave details.

Another method is to make holes on the bamboo sticks and use simple ties to fix them together (table 2.8 example 6). This method alone though is relatively unstable, unless combined with more tensile elements or the ties are made of metal and are combined with fasteners.

Often the sticks are simply fixed together with pieces of bamboo that interlock in the core of the joint and are fixed on their ends not to slide (table 2.8 example 8).

The use of fasteners is also common (Ubidia, 2015, p. 43). Nails are used both for lap and linear joints, as shown in figures 2.65 and 2.66. They are often used in truss-like structures, when axial and shear forces are mainly important for the assembly. On the other hand, bolts could alternatively be used and are actually preferred in minimal bamboo structures (Ngoma, 2004). For example, when the structure relies on beams composed by one single bamboo stick with limited amount of supports, bolts can be more effective, especially for corner joints or joints that are far from the base of the structure, for which bending and torsion can be very critical issues.

For bolted connections, local fillings with concrete are sometimes used as a reinforcement method (table 2.8, 12). These can be realized simply between the internal cavities of the bamboo (H. Garecht, 2010).

Further than the basic joining methods that are described above, there are many more elaborated connection details which aim to improved performances and strength. A simple example that is based on a realized pavilion built in Shangai in 2007, by MUDI Architects (Minke, 2012), appears in figure 2.66 (right). Further examples of more advanced steel joints are presented later in this chapter, in the category of 'hybrid joints in construction details'.

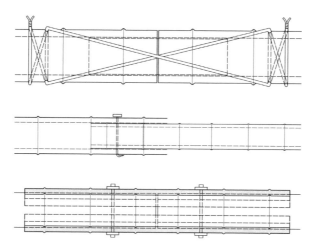

Figure 2.65 Three examples of linear assemblies, (top) tightened with ropes, (middle) nailed together, (bottom) bolted together with halved bamboo sticks.

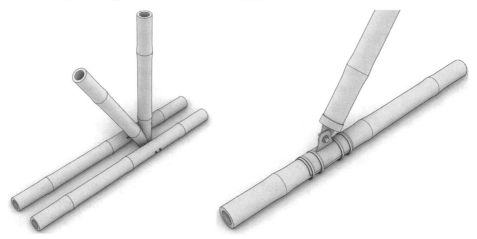

Figure 2.66 (Left) Implementation of dowels on multi-axial lap joints (screws or bolts are the most common choice in such cases), (right) Intermediate hinge-connector made of steel. A steel profile is attached on the bamboo with pressure ties. A cap is fixed on the end of the adjacent bamboo beam element. On both steel surfaces a flap is welded. Finally, the two flaps are bolted together, forming this way a hinged joint.

Form-locking

Proper form-fitting is essential, especially in traditional structures that are built exclusively with bamboo, without metal reinforcements. There is a wide variety of form-locking joints, often combined with interlocks that involve dowels made of bamboo, as it is mentioned in the paragraph about 'force-closure' joining techniques. Some of the principle methods to treat bamboo at the areas of the joints, such as cutting and splicing are presented in relevant manuals (Lopez, 1981b, pp. 7-9) as well as explicit assemblies that are widely implemented in modern projects (Lopez, 1981b, pp. 21-23).

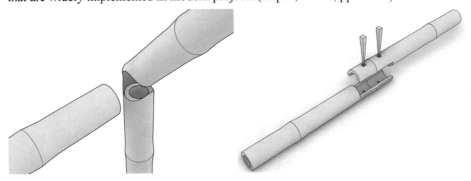

Figure 2.67 (Left) Butt bent joint, (right) Linear form-locking stick joint.

A fundamental difference between bamboo and paper-tubes is related to the direction of the fibers within the component. In the case of form-locking joints, the fiber-orientation in bamboo is quite beneficial, especially when the joints are mostly subjected to tensile forces. On the other hand, the majority of industrially produced paper-tubes is formed with a winding process and so the paper-sheets are laminated on top of each other with a certain angle. Considering the seams created within the tube as a result of this process, creating cut-outs on the tube could affect significantly its structural performance.

Hybrid joints in construction details

A variety of connections is presented that combine the joining principles described above, in highly functional designs. These include multi-axial nodes that involve wood and steel elements in the assembly and fundament details with concrete and steel fittings.

- Nodes

A rich variety of designs is hereby discussed. A common problem that is tackled in a few different ways, is to ensure a stable fixation between the main connection elements and the bamboo. Tolerances from the natural material need to be absorbed by the design.

A plug made of natural wood is the first example of node. The bamboo pipes are plugged-in and fixed in terms of adhesion. The endings of the bamboo are slightly thinned to fit in the node. The node is fixed with a mixture of concrete, used to fill the gap between the wood and the bamboo. The ring shaped close to the edge of the bamboo, as shown in fig. 2.68, helps to form a more stable fixation, when the concrete solidifies, as it forms interlocks on both sides. This connection has been studied and implemented by 'Bambutec-Technology'.

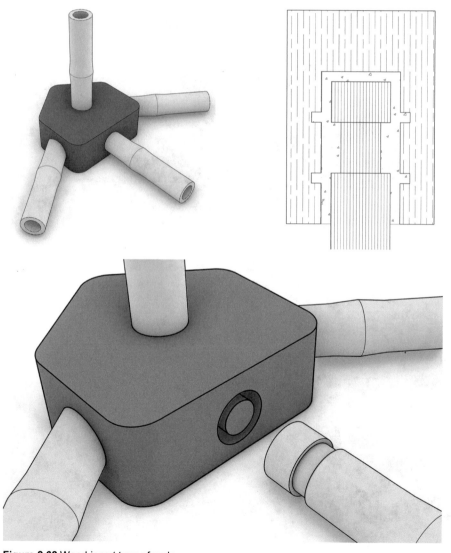

Figure 2.68 Wood-insert type of node.

In the category of steel nodes, a 'gusset node' is the most basic design. This intermediate connector is composed of welded slits that are fixed on the bamboo pipes with pressure rings. The details for the connector as presented in fig. 2.69 are based on a prototype developed in a workshop under the name of architect Renzo Piano (Yu, 2007, p. 114). The limited contact area between the node and the bamboo is the main disadvantage of this connection that makes it inefficient for the case of paper-based tubes.

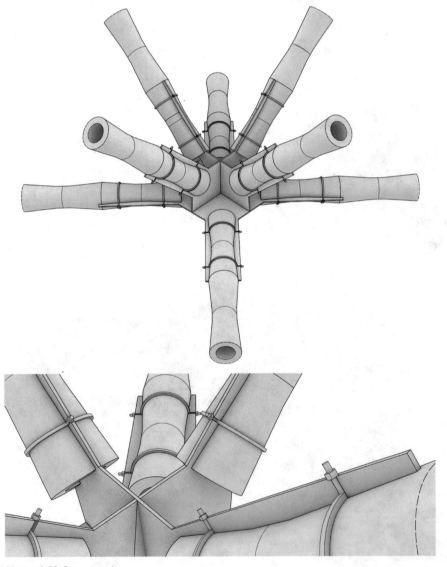

Figure 2.69 Gusset node

A design that maximizes the contact area, between the node and the bamboo, is shown in fig. 2.70. Steel pipes are used to reinforce the bamboo at the area of the joint and stabilize the vertical steel plates together with the bamboo. A multi-axial connector, similar to the design shown in fig. 2.42, in paragraph 2.2.4.1, is placed in the center of the assembly. The different elements are bolted together. This joining method is studied by architect Shoei Yoh (Yu, 2007, p. 114).

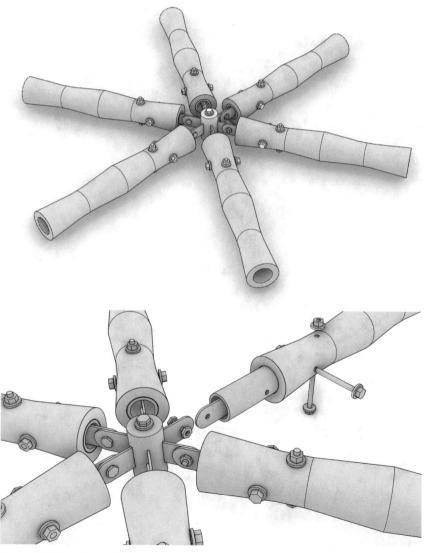

Figure 2.70 Plug-in steel joint

A design that merges the previous solutions is presented in fig. 2.71. Steel plates are casted in concrete, in the bamboo. Pressure ties are used to reinforce this part of the joint. The plates are hinged together at the center.

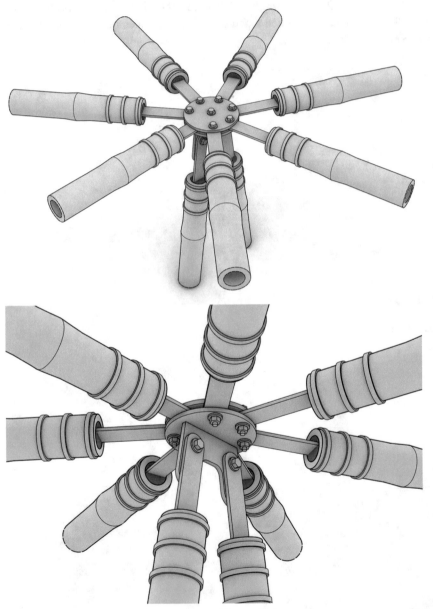

Figure 2.71 These details are based on the reference project 'Bamboo Pavilion', built in Shanghai (Heinsdorff, 2010).

A high-tech solution is presented in fig. 2.72. The edges of the bamboo are treated to get to a conic shape (concrete filling for rigidity is possible) and be fitted in the sleeves made of steel. Then all sleeves are plugged within the central multi-plug component. This design does require a more advanced manufacturing process, comparing to the previous ones. A point for discussion is if such an elaborate solution is the right approach for short-term installations and what happens with potential reuse of the joints.

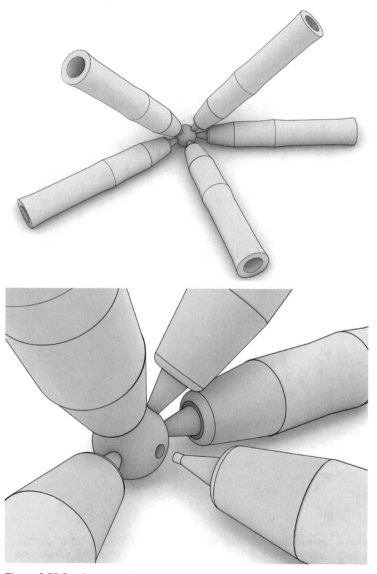

Figure 2.72 Steel truss-node, studies developed by Christoph Tönges (Minke, 2012, p. 47).

- Fundament details

Three simple methods of mounting a bamboo structure on the ground or a base were identified within the review process and are presented hereby. Concrete is often used to bind the bamboo with the joining elements together. It also makes easy the integration of steel bars, to reinforce the fundament (fig. 2.73, top). The downside of this concept is that the joint is not reversible.

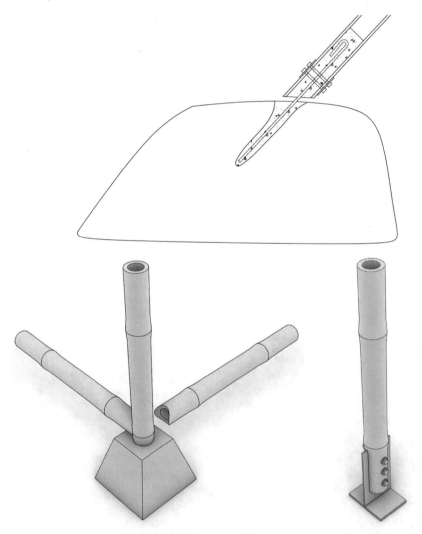

Figure 2.73 Fundaments for bamboo structures. (Top) anchored in stone (Enrique, 2016), (bottom left) casted in concrete-foot, (Natural-Homes, p. 15), (bottom right) steel foot casted in the bamboo with concrete.

Summary

The observations regard the applicability of joining techniques for bamboo on structures composed of paper-tubes. All in all, some of these techniques are already identified in realized paper-structures, such as wooden and steel nodes. Below, the different concepts are discussed one by one.

Designs of dry form-locking joints, like these presented in fig. 2.67, could damage the paper-tubes, especially because of the manufacturing process that leads to seams along the length of the tube. Joints that combine form-locking and adhesion could be effective for paper-tubes, especially when combined with fiber reinforcement. The recyclability of such assemblies is a crucial aspect to define, based on the type of adhesive. As in principle paper-tubes, when uncoated, are recyclable, the recyclability of joints would be the challenge. These could be cut-out and processed separately within the recycling process.

The weaving methods are not expected to be as effective for paper-tubes. The natural texture of bamboo, with the ribs and the rough surface, is beneficial for lap joints secured with knots, as this texture increases friction. The skin of bamboo is significantly harder than that of paper tubes, for which high friction from tightened ropes could cause tearing. Thus, such a joint would not be durable.

The placement of fasteners (nails, bolts, stick joint) can weaken both bamboo and paper-tubes, especially for shear and bending -in the direction of the length-. Thus, simple mechanical fixations, without reinforcement, could only work for basic assemblies, such as continuous walls (2.2.4.1, fig. 2.27), flooring systems etc. In principle, successful implementation of fasteners would require optimal form-fitting as a basis. Next to this, local reinforcement is often required, such as metal plates, to hold the elements under pressure and prevent enlargement of the holes.

A variety of nodes has been presented. Taking into consideration the main criteria of simplicity in assembly, stability and reversibility of the joint, the steel nodes presented in fig. 2.69 and 2.70 present clear advantages. On the other hand, focusing on the aspect of stability, the concrete bond between the joint (wooden or steel plug) and the bamboo in the nodes of figures 2.68, 2.71 and 2.72, absorbs tolerances and increases the stiffness. To cast concrete in a paper-tube, it would be recommended to first apply a layer for water-proofing. The node of figure 2.72 requires the most intensive production process in terms of consumption of raw material (steel), energy and effort.

Regarding the fundament details, as paper-tubes function best when under compression, full support of the section for optimal transmission of forces through surface is recommended. Hence, the concrete-foot would be preferred. Moreover, joining principles from the nodes could also apply for the fundament, such as the hinged joint with a steel cap, shown in fig. 2.66 (right).

Literature

Cogley, B. (2019). CO-LAB Design Office creates bamboo yoga pavilion in Tulum. Retrieved from https://www.dezeen.com/2019/07/26/luum-temple-co-lab-design-bamboo-yoga-pavilion-tulum/?li_source=LI&li_medium=bottom_block_1

Enrique, V. (2016). The technology transfer systems in communities, product versus processes. *ScienceDirect, Elsevier, Procedia Engineering 145, International Conference on Sustainable Design, Engineering and Construction*(145), 364 - 371.

Gumangan, N. Bamboo Design and Construction in the Philippines: the Cabiokid Experience. Retrieved from https://worldbamboo.net/wbcix/presentation/Gumangan,%20Nars.pdf

H. Garecht, J. S., A. Ott, J. Franz, M. Heinsdorff. (2010). Steigerung der Tragfähigkeit der Knotenanschlüsse der Bambusrohre für das DeutschßChinesische Haus der EXPO 2010 durch Optimierung des BambusßBetonßVerbundes.

Heinsdorff, M. (2010). Bamboo Pavilion for the Expo Shanghai. *DETAIL*, 1046-1052.

Kaminski, S. (2016). Structural use of bamboo: Part 1: Introduction to bamboo. *Structural Engineer*.

Lopez, O. H. (1981a). *Manual de Construccion con Bambu*. Retrieved from

Lopez, O. H. (1981b). *Manual de construcción con Bambú* (E. T. C. Ltda Ed.): Universidad Nacional de Colombia.

Minke, G. (2012). *Building with Babmboo. Design and technology of a sustainable Architecture.*: Birkhauser Architecture.

Natural-Homes. Bamboo in construction. Retrieved from http://naturalhomes.org/img/bamboo-in-construction.pdf

Ngoma, I. (2004). Sustainable African housing through traditional techniques and materials: a proposal for a light seismic roof. . *13th World Conference on Earthquake Engineering, Vancouver, B.C., Canada*(Paper No. 170).

Raj, A. D. (2014). Bamboo as a Building Material. *Journal of Civil Engineering and Environmental Technology, 1*, 56-61.

Ubidia, J. M. (2015). *Manual de Construcción con Bambú*: Red Internacional de Bambú y Ratán, INBAR.

Xiao, Y. (2019). *Research development of glued laminated bamboo (glubam) and cross-laminated bamboo and timber*. Paper presented at the 3rd International Conference on Bio-Based Building Materials, Belfast, UK.

Yu, X. (2007). *Bamboo: Structure and Culture*. Retrieved from

2.3.3 Textiles and composites

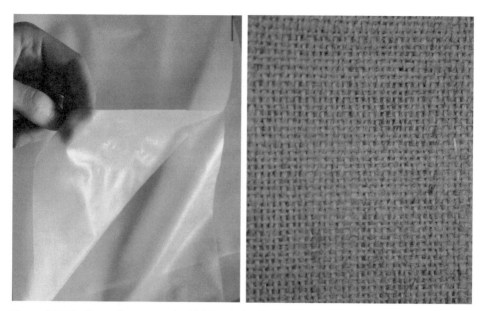

Figure 2.74 (Left) membrane sample, (right) sample of jute textile

About the materials

Textiles are viewed as a useful reference when brainstorming about building with paper. The fact that both kinds of materials are basically produced in flexible sheets and commonly present good tensile properties in the machine direction are perhaps the most obvious reasons for this. Looking into using paper in assemblies, in a similar way to woven textiles and extruded membranes, there is a lot to learn from the joining techniques applied in that field.

The kind of composites that are more interesting in relation with paper-based products are these that combine natural textiles with membrane-like materials. In addition, the principle idea of using textiles to reinforce paper-based products for construction purposes and membranes to achieve weatherproofness is a concept in discussion by researchers in this field. In this spirit, these two categories of materials are hereby merged within a common paragraph. In paragraph 4.2.2 a case study of a connection with reference to this cross-link is presented. Furthermore, composites, in the broad sense, are used to design thin-shell components and connectors for lightweight structures. A few examples in this area are addressed that trigger interesting points for discussion.

2.3.3.1 Textiles

Textiles are commonly used for the erection of temporary or mobile structures, due to their flexibility, potential for collapsible assemblies, light weight and ability to resist wind. Even though in technical terms there are no standards for textile structures, it is a field that is advancing constantly. Like paper, in principle, textiles are most commonly anisotropic materials that present high tensile strength along the machine-direction. The density, strength and orientation of the yarns that are weaved together play an important role for the quality of the final product. The origins of textile products can be animal, plant, mineral or synthetic -based. For construction mostly the last two categories are preferred due to advantages in durability, weather resistance and cost, despite the ecological issues. Table 2.9 presents some of the most common joining principles used to bind, fix or tension textiles that are afterwards discussed in detail.

Table 2.9 A selection of joining techniques applied in textile structures.

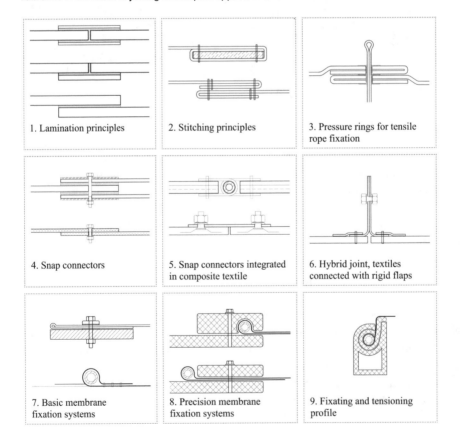

1. Lamination principles	2. Stitching principles	3. Pressure rings for tensile rope fixation
4. Snap connectors	5. Snap connectors integrated in composite textile	6. Hybrid joint, textiles connected with rigid flaps
7. Basic membrane fixation systems	8. Precision membrane fixation systems	9. Fixating and tensioning profile

Material Closure

Welding is a common process used to join membranes, PVC coated textiles and some other types of synthetic textiles but with limited results in terms of bond strength. Perhaps it could be used to bind paper products that are coated with a synthetic film. By aligning the two pieces of textile so that they share a contact line and applying heat and moderate pressure the textiles bond.

Lamination is very commonly used to create a bond that is not water or air permeable, especially when the desired assembly presents a geometry that is not easy to realize with common welding equipment. Some examples of application encounter inflatable mattresses and boats. Some of the most popular types of adhesives are PVA, PUR, HMA, rubber-based, epoxy-based adhesives. Starch-based adhesives are often used within the production of some types of textiles but not for high-strength bonding.

An alternative to adhesives is tape (single or double-sided) and is often used as a fast way of joining textiles, due to the fact that it does not require time for drying. Combination with stitches to prevent delamination is possible.

Force closure

The most known technique is stitching. The textile is often single-folded along the joining line, to increase strength. Next to this, it is possible to use special lining to improve stability. A variety of stitching methods and tools apply depending on the boundary conditions. Custom designed dense stitching patterns are also used to reinforce a surface. Next to stitching, velcro bands are also worth-mentioning, as a popular way to create operable seams. The two elements that stick together are attached on the adjacent textiles along the perimeter and diagonals. All techniques that involve stitching require experimentation to define which type of paper-sheet, yarn and pattern would be more effective.

Figure 2.75 (Left) Prototyping with textiles, (right) basic connectors in everyday fabric products

Common mechanical fixations refer to a variety of snap connectors, such as rivets, clinches (Knippers, 2010, fig. C 6.38, p.118), pressure rings and a variety of button-like connectors. All these can be attached on the textiles with simple tools. These could also be attached on paperboard, as mentioned in paragraph 2.2.2, table 2.4.

In small scale structures, often the areas with mechanical fixations are reinforced with patches of stronger textiles (stitched, laminated or welded). The same concept is often used to integrate tensile cables. For the same reason, in larger scale construction, thin plates of rigid materials are often integrated in the areas of the joints (Knippers, 2010, p. 182).

Figure 2.76 (Left) Steel hinge fixed on composite (PVC-coated polyester) suspended roof structure (Olympic stadium, Munich, DE), (right) flexible membrane tensile roof-structure (Rotterdam, NL).

Figure 2.77 'The wing shelter' design made by a group of students, 2014, Buck Lab studio (Fall 2014, TU Delft, direct supervisor: Jerzy Latka, head of studio: Marcel Bilow).

When thinking about textiles in construction, tensile structures are perhaps the most memorable example. In figures 2.76 and 2.77 three different thin-shell tensile structures are presented. Cables, welded seams and woven stripes are used to keep the surface of the shell in shape, depending on the thickness and quality of the surface-material. The 'wing-shelter' is an inspiring study about how paper could be integrated in a double-curved surface, by using its tensile strength and protecting it from wrinkling and tearing because of twisting. The woven surfaces are mounted on paper-tubes along the perimeter (joining principle 7, table 2.9), by creating simple loops around the paper-tubes and fixing glued lap joints.

Form-locking

In the case of textiles form-locking joints are commonly presented as looped or threaded knots. This way, simple knots formed between the edges of textiles with split-ends is a good example of basic form-locking. In a different way, thicker textiles, like felt, can be cut intro puzzle fittings or other special form-fitting designs. Such principles are often used to form organic and parametric geometries for interior design, art installations etc.

Hybrid solutions

In table 2.9, details 5 and 6 present examples of how rigid plates can be integrated at the areas of the joints in combination with fasteners and stitches if needed, either by creating a flap or in a linear assembly. Moreover, examples 7 to 9 show common details of clamps that are used to secure the textile and create tension. Similar examples can be seen in the 'Atlas Kunststoff + Membranen' (Knippers, 2010, p. 180).

In figure 2.78 the design concept for a collapsible pneumatic structure, supported by a bimini frame is shown. There, joining technique 7 from table 2.9 is used in many areas of the design, to connect the fabric skin with the bimini.

Figure 2.78 'The pop-up office', examining inflatable structures. MSc thesis, TU Delft, 2016

2.3.3.2 Synthetic composites

As some multi-layered paper products are often characterized as composites, due to the integration of polymer films in the matrix, this material category presents great interest.

Synthetic composites are manufactured for targeted applications and so the basis materials are configured to obtain optimum results. When the application is structural, the components can be optimized for the specific purpose, by alterations in the matrix of the composite. To this respect, synthetic composites are often reinforced with fibers and adhesives. The selection of these has great impact on the recyclability of the end product.

This way, this category of materials includes a wide range of products. Composites are often used in small-scale structures, where lightness is important, such as vehicles or other thin-shell structures.

In this paragraph principle joining methods are discussed and some examples from experimental studies, in the area of design and construction that present ideas for the joining techniques are demonstrated.

Table 2.10 A selection of joining techniques applied on composite materials.

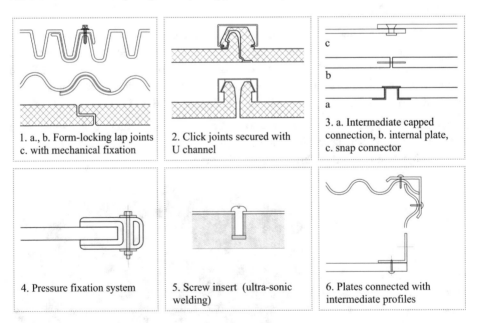

| 1. a., b. Form-locking lap joints c. with mechanical fixation | 2. Click joints secured with U channel | 3. a. Intermediate capped connection, b. internal plate, c. snap connector |
| 4. Pressure fixation system | 5. Screw insert (ultra-sonic welding) | 6. Plates connected with intermediate profiles |

Material closure

The adhesives used are in principle the same as in the previous paragraph about textiles. Epoxy resin is a common choice of thermoset used to manufacture strong composites. It is also often used to glue components together. There is a variety of such products that are produced with different methods and qualities. In principle epoxy resin is not good for the environment and therefore its use in eco-friendly structures is better avoided.

To increase the strength of the bond fiber reinforcement of the laminate is recommended (Herzog, 2003, p. 153). Fiberglass is often used in this case, though abiding concerns for ecological aspects.

Force closure

In fig. 2.79 (top) an example of implementation of bolts with rubber-rings, to prevent damages on the material is shown. In general, difficulties in the implementation of fasteners are observed. More specifically, applying sufficient clamping force is one problem, whereas loss of preload due to creep and relaxation are also identified (Murray, 2012, p. 4). This is the reason why joining principles like the one presented in example 4 of table 2.10 are used occasionally, in order to distribute the clamping force on a bigger surface. This way, fastening of composites is a subject in development and custom solutions for the different composite materials apply. Interesting samples are identified in the area of 'threaded fasteners for plastics' (infastech). For the same reasons, when bolts are applied the amount of torque used is crucial, as excessive torque could damage the material. To improve such problems, technologies for embedding steel elements in composites are developed, either within the production process or with adhesives later on. Relevant examples are presented in the 'Atlas Kunststoff + Membranen' (Knippers, 2010, fig. C 6.15, p.112, 170, 163).

An interesting method used to integrate metal components in thermoplastic parts is ultrasonic welding (DUKANE Guide, 2011).

Fiber reinforcement of composites can advance the structural performances and distribution of stresses (Schürmann, 2005). A relevant study about the implementation of pinned and riveted fasteners on fiber reinforced plastics suggests realignment of the fibers at the areas of the joints to improve the distribution of stresses, inspired by the organic shape of fibers within natural wood, around notches (Holger Seidlitz, 2017).

Form-locking

Simple lap joints that rely on form-locking in combination with mechanical fasteners is one option, as shown in example 1 of table 2.10. It is commonly used for thin-wall materials. A more elaborate joining method, shown in example 2, is 'click' joints. The 'click' system can either be formed directly between the two elements or with a third element, as indicated in the example.

Figure 2.79 Bolted joints with fiber rings hold the panels fixed together. The components were produced by casting epoxy resin, reinforced with fiberglass, within the boundaries of the MDF integrated structure and the support of a special adjustable mold that was used as a base for the production of all the panels (archive: 'The Thing', team-project, Design Informatics studio, TU Delft, 2015).

A design suggestion of a cap-node for fixing triangular panels in a lightweight dome structure that combines the principles of form-locking (multiple cavities) and force-closure (central bolt) is shown in fig. 2.80.

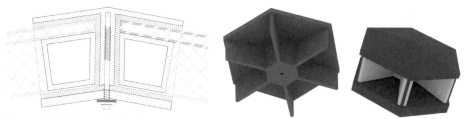

Figure 2.80 'Extreme Project', a study on a shelter for mountain Everest, camp 3, TU Delft, 2014. A detail of a node for a dome structure, in carbon fiber.

Summary

Hereby the potential implementation of joining techniques from textiles and composites on paper-based assemblies is summarized in two separate sections.

- Textiles

All in all, the majority of the joining techniques for textiles presented above could also be used for thin paper-based products.

Of course, overlapping is identified in adhesive joints. An interesting point is how could the implementation of polymeric films on paper create potential for welding sheets into one skin.

To use stitching as a joining technique for paper-based assemblies, experimentation is required to identify effective stitching patterns. To prevent tearing between the stitches, thick sheets and non-dense stitching would be a good point to start from.

In general, for thin paper materials, joining principles that divide stresses along a surface or a line are preferable comparing to single connectors that lead to point loads. In the case of punctual connectors, this would mean that depending on the material characteristics (thickness and matrix) and the requirements, the distancing between the connectors could be examined.

Velcro connectors are not expected to be effective, especially if the tensile forces exceed the capacity of the bond between velcro and paper. Combination of glue and stitches to attach the velcro on the paper-product could lead to better results.

Examples regarding the implementation of local reinforcements – flexible patches or rigid plates – at the areas of the joints, with the aim to optimize the performance of connectors such as snap-fits etc. could also be transferred to paper-based structures.

- Synthetic composites

The use of epoxy resin as an adhesive is a method that applies for paper-based products as well. However, it affects immensely the recyclability of the components where it is applied.

Fiber reinforcement between the different layers of an assembly to improve the resistance of the joint formed between them against shearing is a concept with potential. Next to this, the idea of laminating fibers by imitating patterns from natural materials (Holger Seidlitz, 2017), such as wood notches, to surround the hole of the fastener, is an interesting concept, but it would require significant effort in terms of design, experimentation and production.

The investigations on embedment of fasteners in the structural elements either in the production process or later with adhesives, as described above (force-closure), could improve the performance and durability of the entire structure. On the other hand, any steps that need to be integrated in the production process and require high precision increase the complexity of a project and also the costs.

Thus, the concept of ultra-sonic welding gains interest, even though technically it is not possible to implement on pure paper, without incorporating an amount of plastic in the component, at least exclusively in the area of the joint. When hollow plates, such as honeycomb boards, are part of the layering, then some of the cavities could perhaps be filled with a polymer suitable for this process.

For plate or shell structures, where the structural elements are primarily subjected to compression, when the implementation of mechanical fixations presents significant drawbacks, using intermediate connectors, such as the example of figure 2.80 is an alternative solution.

Literature

Herzog, T. (2003). *Holzbau Atlas*. Basel, Boston, Berlin: Birkhäuser Edition Detail.

Holger Seidlitz, F. K., Nikolas Tsombanis. (2017). Advanced joining technology for the production of highly stressable lightweight structures, with fiber-reinforced plastics and metal. *Technologies for Lightweight Structures*(Special issue: 3rd International MERGE Technologies Conference (IMTC)).

infastech. Threaded Fasteners for Plastics. In.

Knippers, J. (2010). *Atlas Kunststoff + Membranen (Detail Atlas)*: DETAIL.

Murray, W. M. (2012). Some Curious Unresolved Problems, Speculations, and Advances in Mechanical Fastening.

Schürmann, H. (2005). *Konstruiren mit Faser-Kunstoff-Verbunden*: Springer Berlin Heidelberg New York.

2.3.4 Steel

Figure 2.81 Eiffel Tower, Paris (FRA)

About the material

In principle, like paper, steel presents high strength comparing to its weight and particularly high tensile strength. A fundamental difference between steel and paper-products that challenges the discussion about transferable solutions for the areas of the joints is related to the strength of the basis material. In general, within an assembly made of steel, when a joint is tested, the failure is expected to appear on the side of the joint, for example the welding line or the fasteners. On the other hand, in a paper-based assembly, mainly the opposite is expected to happen. It is interesting to examine some possibilities through this filter.

Steel is a high-performance isotropic material used to build lightweight construction that corresponds with the highest structural requirements as well as durability. Its production requires non-renewable raw materials and is highly energy intensive. In contrast, it is a highly recycled material. A few of the major applications include skeletal and truss structures with large spanning, façade systems and reinforcement of concrete. Steel is often used to reinforce hybrid structural systems. Some examples that are in closer relation with this research regard simple mechanical fixations and also customized intermediate connections (see figure 2.45).

The available production and post-processing methods create a wide spectrum of possibilities for the forming of special elements, also joining elements. Steel structures present a highly advanced technological background as well as a variety of joining methods that are important to discuss within this research.

Table 2.11 Review of selected joining and assembly principles for steel.

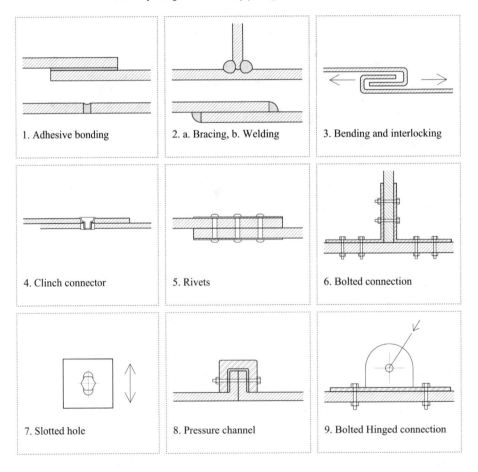

1. Adhesive bonding	2. a. Bracing, b. Welding	3. Bending and interlocking
4. Clinch connector	5. Rivets	6. Bolted connection
7. Slotted hole	8. Pressure channel	9. Bolted Hinged connection

Joining techniques

The joining techniques for steel structures are extensively described in EN (1993-1-8, 2005). There, joints are mainly divided in two groups: bolts, rivets or pins and welded. The first category is hereby identified as force-closure and the second as material closure, whereas the form-locking methods refer to the possibility of integrating parts of the joint in the main components.

Material Closure

Welding (table 2.11, 2) might be one of the most common techniques preferred to join steel and is recommended for shear forces. It is used to join two elements along a line, by fixing them in place, locally heating the material until steel reaches its melting point and when it cools down and rigidizes again, the two elements share a common edge. This technique could not be transferred such as to paper-based structures, as to redirect the fibers not only heat, but also pressure and presence of water are needed. However, in the same spirit, couching, does create a similar effect, even though the conditions to perform this technique are entirely different, as explained in paragraph 2.2.2.

Additionally, following the technology of 3d printing with pulp-based mixtures that is currently in development, it is a possibility to use the same paste as an adhesive, in order to join paper-based materials. However, this technique would be much closer to gluing rather than welding.

Adhesives are sometimes used instead of welding to speed up the assembly process and occasionally preferred over fasteners to prevent corrosion (for permanent joints). The main types of adhesives for steel identified as epoxy adhesives and polyurethanes, both present products with very high strength but also significant ecological downsides.

Force closure

The main types of fasteners encountered here are bolts and rivets.

In EN1993-1-8 (paragraph 3.4) a set of rules is provided for different cases such as for excessive shear forces, slipping, tension and specific types of fasteners are recommended per case.

In principle, bolts are preferred for tension and preloaded bolts for assemblies with optimized performances at the 'ultimate limit state'. In relation with this research, injection bolts could be considered to enhance performances (resin is added to create a thin wall between the fastener and the adjacent elements) and prevent tearing of the paper-based components.

Some advantages of rivets are that they can be fixed within very thin materials (whereas for bolts this could be problematic) and also that vibrations don't make them loose. On the other hand, they cannot reach the amount of pressure created by bolts. Next to this, they cannot be reused.

Regarding the dimensioning of mechanical fixations, as the standards provide detailed rules, it is sensible to consider that for paper-based assemblies mostly higher distances shall be attempted.

A relatively modern joining method for steel that aims for rapid assembly is clinching. It works well for shearing but not tension. It can be realized with or without local incision. Without incision, no further elements are used, the components are locally deformed, in order to create an interlock that is a result of positive - negative form. In this case the joint formed is non-demountable. As reinforcement methods, combination with adhesion or rivets is possible. This method, for the time being, is often implemented in industrial areas such as automotive, to join thin sheets of steel (KAŠČÁK, 2014).

When special self-clinching connectors are implemented (PEM), the two sheets are punched together and then the connector is applied (table 2.11, 4). The main concept is already known from connectors applied on textile products and membrane structures. Both concepts present potential for a spectrum of paper-based materials (mainly paperboard products), for applications that require low strength for the joint itself.

Form-locking

As steel structures are lightweight and composed of thin-wall elements, the typical form-locking joints as in timber structures do not apply. For example, the first category of clinching without incision that is referred under 'force-closure' is a kind of form-locking joint.

Thinking in a bigger scale, the possibility to bend or press steel in various shapes creates potential for the design of form-locking joints to assemble plates together (table 2.11, 3). This concept is interesting for the assembly of solid paperboard plates, but the forming processes face significant issues related to damages occurring in the material, especially for thicker boards.

Moreover, there is a wide spectrum of steel elements that use the principle of form-locking, but these are merely industrially produced connectors and fittings (bolts, pins, pipe fittings etc.)

Hybrid joints in construction details

Examples of different levels of complexity show how thin plates and hollow profiles are joined and assembled to form lightweight supportive frames.

Figure 2.82 Roof structure built with a combination of welded and riveted profile (De Pont Museum of Contemporary Art, NL).

Some steel-reinforced multi-axial nodes are already presented in timber and bamboo details. A different example of a steel node that clamps all adjacent profiles together is presented in fig. 2.83.

Figure 2.83 Steel joint for parametric façade design (Louis Vuitton Foundation, Paris (FRA)

Summary

In the category of 'material closure' the method of welding elements together is discussed. Based on the material structure of paper-products this method is in general not transferable. A concept that reminds of welding is considered in paragraph 2.2.2. It is called couching and it is about merging sheets of paper in conditions of high humidity and heat. However, this method cannot be applied on thick paperboard or other paper-based components. Moreover, the idea of using pulp-paste, known from research efforts in the direction of 3d-printing, to adhere components together is another possibility that reminds of the welding concept. This solution is in a conceptual phase, as there are issues with the curing process, related to shrinkage and deformations.

Regarding the mechanical fixations, the techniques with best potential seem to be bolts, particularly injection bolts and rivets. Clinch connectors could be tried in paper-based assemblies, though issues due to limited anchoring would be expected.

Considerations in relation with the dimensioning of such joints that involve mechanical fixations, for different occasions, are expressed. As a global observation, in principle, when fasteners are applied on steel, it is more likely that the strength of the fasteners will define the strength of the joint, whereas on paper-based materials the opposite happens. Hence, the design and dimensioning of joints and failure modes would differ significantly.

Finally, deformation of paper-based elements to create form-locking joining techniques, for example between plates is pointed out as a possibility.

Literature

1993-1-8, E. (2005). Eurocode 3: Design of steel structures - Part 1-8: Design of joints. In.

KAŠČÁK. (2014). *Clinching as innovative method of steel sheets joining and a tool of increasing competitiveness of companies in automotive industry.* Paper presented at the The 17th International Scientific Conference. Trends and Innovative Approaches in Business Processes.

PEM. The self-clinching fastener handbook. Retrieved from https://www.pemnet.com/fastening_products/pdf/Handbook.pdf

2.4 Global overview of joining techniques for paper structures

Introduction

In chapter 2 the subject of joining techniques for paper-based assemblies and structures has been reviewed. The main focus is on presenting the 'state of the art' and identifying potential influences from conventional building methods for the advancement of the research on this subject. The main goal is to create an informative collection of joining techniques, explain the pros and cons of these and discuss their functionality while indicating potential applications. The first part of the review is analyzed in the three sections of basic joining techniques, techniques specialized on lightweight paper-based boards and techniques specialized on paperboard profiles.

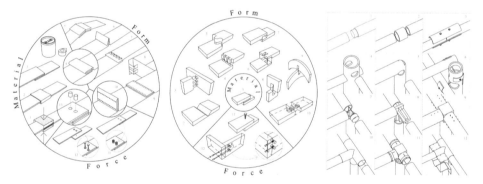

Figure 2.84 Joining principles for paper-based products and assemblies, as presented in tables 2.4, 2.5 and 2.6.

In the second part of the review joining techniques implemented on design and construction elements made of timber, bamboo, textile, composite and steel products are analyzed and their applicability on paper-based assemblies is discussed. Hereby the outcomes from the review process are briefly summarized per material category (2.4.1). Then the global overview is presented (2.4.2), with the aim to outline the greater potential of the review process for the development of further studies and research. Moreover, the influence of the review process on the next phase of the research, that is presented in chapters 3 and 4, where specific joints are investigated, is discussed in the 'epilogue'.

2.4.1 Summary of findings per material category

Hereby the observations from the review process are summarized, as an introduction to the global overview of joining techniques presented in 2.4.2. As there is a wide variety of paper-products it is important to notice that the different joining techniques discussed are attributed to specific products per case. As the focal point in this research is skeletal structures composed of paper-tubes, further attention is drawn to this area.

Paper-based assemblies

Overall, as explained in paragraph 2.1 the 'state of the art' is based on knowledge derived from the paper-manufacturing industry, a collection of reference projects and studies and draft experiments performed for this research.

Lamination (2.2.2 Basic joining techniques) is a technique best explored by the paper industry, where adhesives such as starch, dextrin, water-glass, PU and PVA glue are used in different occasions, such as the production of multi-layered boards, gluing packaging products, labelling etc. From the adhesives mentioned above, PVA glue is commonly used in timber connections and a variety of products is available. Draft tensile tests on laminated lap-joints formed with kraft paperboard and 'Ponal' glue (PVA), performed in this research, showed failure on the side of the paperboard and indicate potential relation between the fiber orientation in the paperboard and the behaviour of the laminate at ultimate limit state (paragraph 2.2.2, graphs 2.1 and 2.2). Starch-based adhesives are known as a more ecologically friendly solution, but often present lower strength.

Within the category of force-closure stitching and stapling are basic methods used to attach paper products with limited thickness, especially in the process of book-binding and packaging. Form-locking mechanisms are again identified mainly in the field of packaging products, commonly as creased edges and flaps used to assemble 3d-objects from thin sheets and boards.

Regarding techniques specialized on boards (2.2.3), 3d-puzzle assemblies are often used to create lightweight designs, whereas modern fasteners such as screws with wide spiral are used to attach honeycomb boards (fig. 2.14).

The techniques discussed above, both for force-closure and form-locking present, in principle, limited structural performance. Therefore, assemblies that are closer to the field of construction regard hybrid solutions found in realized structures, such as the 'Nemunoki Children's Museum' (McQuaid, 2003), the 'Apeldoorn Temporary theatre' (TU-Delft, 2008) and the 'Wikkelhouse' (Fiction-Factory). These solutions are presented in detail in the section 'construction details' of paragraph 2.2.3.

In the separate category of joining techniques for paper-based profiles (2.2.4), assemblies that involve round paper-tubes present the greatest variety of designs, based on the review of realized projects (2.1.1). These regard assemblies built based on form-locking similar to bamboo structures fixed with ties, force-closure realized with bolted joints reinforced with pressure profiles and a variety of hybrid solutions. The last ones include mainly: light and massive wooden nodes (McQuaid, 2003), steel nodes (Octatube) and experimental 3d-printed nodes. Further details and references are provided in paragraph 2.2.4.1 'Joining techniques for paper-based tubes'.

Timber

From the field of timber structures (2.3.1, table 2.7) practices on lamination are particularly useful, especially when looking into laminated timber and CLT construction. Based on experimental prototypes, the use of PVA wood glue seems to be a good solution for laminating paper-based products for construction purposes. Based on draft prototyping experiments, it adheres well, without being absorbed by the material in depth, in which case it would affect the recyclability of the connected elements. Draft prototypes have shown that staples, nails, screws and bolts often suit also for paper-based assemblies. Staples are effective on paperboard and multi-layered corrugated board (maximum joint thickness 14mm). Nails could work to prevent shearing but not for tension or compression as the fasteners would fall-out due to poor anchoring. Screws present limited performance as well due to poor anchoring. Bolts are more effective, but moderate torque is required, to prevent material damage, as paper-products are in principle soft. Even though, especially in the case of screws and bolts, the failure modes are likely to differ, the guidelines provided in EN1995-1-1: Section 8 – Connections (Leijten) are a useful base. Joining techniques that rely on form-locking (carpentry joints) present significant issues if applied on paper-based products, depending on the matrix -layering- of the assembly.

Bamboo

Bamboo structures (2.3.2, table 2.8) present assemblies that are closely related to examples of paper-tube structures. Joints that combine form-locking and gluing (fig. 2.64, 2.67), in combination with ties (fig. 2.65), to prevent failure due to tensile forces, could be compatible. However, cut-outs on the tubes could weaken the profiles significantly. Fasteners are expected to demonstrate limited performance without local reinforcement, such as pressure profiles (fig. 2.65, 2.66). Examples of intermediated connectors made of wood (fig. 2.68), steel (fig. 2.69 – 2.72) and concrete (fig. 2.73) are also transferable. Designs that offer increased contact surface between the profile and the joint present better potential. For example, the 'gusset node' (fig. 2.69) is less recommended, whereas the plug-in steel node (fig. 2.70) is a better solution. In some designs, like in the multi-hinge node (fig. 2.71) concrete is used to adhere the main components of the connector with the bamboo. This idea could also be worked-out with paper-tubes. Experiences with prototyping showed no obvious damage on the tube, following the casting of concrete directly in it. Coating the tube, in advance, in this case would be recommended. Of course, such a joint is not reversible, a fact with great impact on the recyclability of the assembly.

Textiles

A different case is this of textiles (2.3.3.1, table 2.9) that mainly present solutions for thin materials, thin-shell and tensile structures. Adhesion through welding is a possibility for PVC coated textiles and so could it be for coated paper-sheets. The strength of the welded seam is not expected to be high and it depends on the thickness of the film that covers the paper-sheet. Stitches, combined with double side tape to prevent tearing, is another possibility for lap joints. Mechanical fixations such as rings, rivets, clinches and a variety of snap-connectors are possible to attach on thin paper-sheets and paperboard. Local reinforcement of the material-surface is recommended to improve the strength and durability of the connector. Methods for local reinforcement either with integration of rigid plates or patches are only some ideas in this direction. Looking into tensile structures, flexible or rigid cables are commonly used to carry the loads. Thinking about a tensile structure made of paper, it would be interesting to examine if simple joining methods, such as stitching, could be used to embed these elements on the surface of the material while preventing tearing or wrinkling.

Composites

Adhesives are widely used, especially in the manufacturing process and also to connect elements. The use of green adhesives is more common in manufacturing rather than for connections. The reason for this is that green adhesives commonly lead to lower strength. Thus, it is possible to use them better as a binder, but not for high-strength joints. This way, epoxy resin is often used for high performance composites and strong joints, despite the ecological downsides. Composites (2.2.3.2, table 2.10) present more advanced methods, comparing to textiles, for the embedment of fasteners in components within the manufacturing process. Similar possibilities could be examined for objects produced by casting pulp for example. The feasibility of this idea depends on the requirements for the dimensions, the shape and function of the component, as casting is a process that requires significant preparation and machinery. Another technology, used for example to attach fasteners on composites, is ultra-sonic welding. This technique is not directly applicable and would also require the addition of certain polymers, in considerable amounts, within the assembly. Further studies examine the integration of fiber-reinforcement at the areas of the joints to prevent material damage from punctual joints (Holger Seidlitz, 2017). These developments provide inspiring ideas, even though they are mostly not directly transferable to commercial paper-based products and would be faced better as areas for further research.

Steel

The technique of welding is not possible for paper-products. Couching is a method in the same spirit, addressed in the basic joining techniques (2.2.2), that uses the combination of high humidity and heat to soften paper-sheets and merge them in one body. If these conditions could be applied locally, then this principle could perhaps be used to bind thin sheets. The implementation of fasteners (2.3.4, table 2.11) is in general possible, but assemblies from timber construction provide better insights, following the anisotropy of timber and paper. The possibility to reshape steel sheets by bending or deforming them presents interest. This way interlocking mechanisms between thin sheets can be formed. This idea could be better explored with paperboard (2-5mm thick). The relaxation of the paperboard in time is a factor that is not in favour of this concept.

The review of joining techniques lead to a collection of possibilities. These are visualized in table 2.12. The selection of joining techniques to be researched in detail will be defined in chapter 4, after analysing the construction outlines in chapter 3.

2.4.2 Global overview of joining techniques

The structure of the overview

The overview of joining techniques (table 2.12) both summarizes the 'state of the art' and indicates directions with potential for further development, following the identified crosslinks with common construction materials. Hereby the structure of the overview and the accompanying information are explained.

The overview is organized in the form of a pie in which every piece presents the most recommended findings per material (reviewed in paragraphs 2.2 and 2.3). Additionally, each piece is divided in rings that represent the main categories of joints, material-, form-, and force- closure. The outer ring is used to present hybrid solutions, with higher complexity, as these are known based on existing applications in structures.

The content of the overview is supported by complementary tables (2.13 – 2.16) that provide information with regard to the functionality of each technique, in an attempt to classify the individual characteristics. Overall, the evaluation is qualitative and further than the details described here, the legend provides an explanation for the color-coding system, wherever this applies. The joining techniques that are presented in the tables follow the numbering provided in the overview.

First the 'transferability of forces' is described. For this purpose, the ability of a joining technique to transfer pressure 'P', tension 'T', Shear 'S' and 'Moment' is evaluated. A good reference for such a categorization can be found in a presentation for timber connections (Leijten, , slide 12). For example, joining technique 1 (lamination) is effective for shear forces and pressure but less effective for tension and moments. - The coloring in the box with the number indicates the material group in which this joining technique belongs. This indication is considered helpful, as the tables are organized primarily per fundamental joining method and not per material.

As a second element, the technological means that are required to implement the joining technique are ranked as 'low', 'medium' or 'high'. For the category 'low' basic tools and other equipment (e.g. lamination table) may apply. The category medium 'Med' includes widely available forming methods (e.g. wood-working tools, basic pressure forming etc.) and tools for specialized processing methods. The forming methods considered as advanced – 'High'- are these that require licensed skills for the production process (e.g. steel treatment), time consuming processes which are not yet applicable in large scale and innovative equipment. In continuation, critical aspects for the performance of the joint are briefly addressed with keywords, focusing on identifying the area or the effect that is more likely to lead to failure. Finally, recommended applications are given, in order to both indicate potential, but also imply restrictions.

The main trends expressed by the overview are briefly summarized with a few observations. First of all, as the overview intentionally indicates, adhesion is the fundamental method for assembling paper-based products. To summarize the 'state of the art' for joining techniques in paper structures, the main principles regard folded flaps and rigid interlocks (form-closure 1 - 3), fasteners for solid and hollowed boards (force-closure 1 - 4) and hybrid solutions. The last ones often involve other materials that are used either as a reinforcement for the implementation of fasteners (e.g. joints between panels) or to form joints for paper-tube structures (1 - 5), a category of assemblies that represents a big part of the realized paper structures in total.

Regarding the influences from common construction materials, textiles provide joining techniques for thin sheets such as stitches and integration of punctual fixations with local reinforcement methods (patches, fiber reinforcement, enhanced adhesion etc.). Timber structures offer useful input both for adhesives that are suitable for construction purposes and mechanical fixations (dowels, staples, nails, screws etc.). Practices for reinforcement with integration of steel bars, pressure plates etc. are useful as well to enhance the stability of assemblies. Finally, bamboo structures have great influence on paper-tube structures as the majority of joining techniques are transferable to assemblies composed of paper-tubes. Techniques that rely on pressure and contact are expected to be more effective, whereas these that would lead to significant shear forces in the tubes such as the examples 'form-5' and 'hybrid-8' are less beneficial.

Table 2.12 Overview of Joining Techniques – Map of influences and solutions

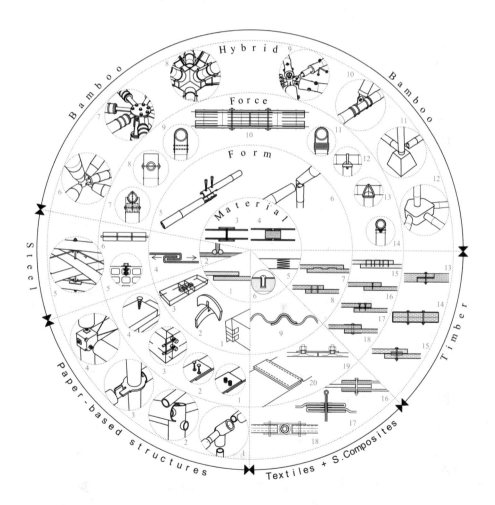

Legend

Material

Paper

Steel

Bamboo

Timber

Textiles & composites

Structural performance (in principle the colour-scale aims to indicate positive characteristics with lighter tones)

High

Medium

Moderate

Poor

Table 2.13 Material -closure (adhesion)

Nr.	Joining Technique	Transferability of Forces				Tech-nology	Critical Area / Aspect	Recommended Application
		P	T	S	M			
1	Parallel lamination					Low	Contact Surface	Fundamental
2	Angular (T) Bonding					Low	Contact line	Non-structural, seal open edges
3	Laminated external fitting					Low	Homogeneity of bond, torsion	Join profiles
4	Laminated in-ternal fitting					Low	Homogeneity of bond, torsion	Join profiles
5	Directly Lam-inated sec-tion					Low	Bending	Join profiles
6	Ultra-sonic welding					High	Contact-surface (polymer/ paper), limited input	Fix elements on paper-based panels

Table 2.14 Form-locking

Nr.	Joining Technique	Transferability of Forces P T S M				Tech-nology	Critical Area / Aspect	Recommended Application
1	Interlocked flaps					Low	Tension on creased areas	Sheets
2	Interlocked plates					Low	Peak stresses at contact areas	Plates and panels
3	Biscuits					Low	Open joint	Panels
4	Bent-locked plates					Low	Precision and durability	Sheets and plates
5	Stick-joint					Low	Peak stresses around locked 'stick' elements	Paper-tubes
6	Bent-locked profiles					Low	Strength of formed tie	Paper-tubes
7	Shear Inter-lock					Low	Limits of chan-nels for shear	Wall panels
8	Dowels					Low	Fall-out (dowel) or tearing (paper-based component)	Plates and panels
9	Interlock-formed plates					Low	-	Cladding

Table 2.15 Force-closure

Nr.	Joining Technique	Transferability of Forces				Tech-nology	Critical Area / Aspect	Recommended Application
		P	T	S	M			
1	Pressure rings					Low	Collapsing of paper (hammering)	Thin sheets/ boards/ origami tensioning
2	Rivets				*1	Low	Fall-out	Thin sheets/ boards
3	Cross dowel					Low	Efficiency and durability depend on base material	Dense layering
4	Screw with wide spiral				*	Low	Fall-out	Honeycomb board
5	Expansion joint				*	Low	Depends on the inner structure of the component	Components with medium sized cavities
6	Pressure rod					Low	Damages at the perimeter of the rod	Butt-joined panels of large size
7	Bolted joint + stability insert (90°)					Low	Failure of paper at contact are with thread	Profiles
8	Bolted cross-interlock (90°)					Low	Failure of paper at contact are with thread	Profiles
9	Bolted pressure-tie (90°)					Low	Failure of paper at contact are with thread	Profiles
10	Bolted joint, with pressure profiles			*	*	Low	Bending moment	Profiles
11	Tightened pressure-tie - 1 (90°)					Low	Weak for moments (vertical axis)	Profiles
12	Tightened pressure-tie - 2 (90°)					Low	Weak for moments (vertical axis)	Profiles

[1] The performance depends on the number of fasteners and the specific arrangement.

13	Tightened pressure-tie - 3 (90°)					Low	Strength limited to the properties of the yarn	Profiles
14	Pinned fish-mouth (90°)					Low	Strength of pin + tightness of knots	Profiles
15	Staples					Low	Fall-out	Boards, max. depth of bond 14mm
16	Nail				*	Low	Fall-out, tearing of paper, overall limited performance	Basic fixation, preferable interior assemblies, or light cladding plates
17	Screw				*	Low	Fall-out	Dense compo-nents
18	Solid rivet				*	Low	Pressure applied on paper, Contact area (rivet-paper)	Solid board
19	Embedded Snap con-nectors					Low	Splitting and dam-age of basis material or connector	Solid elements with limited thick-ness
20	Stitching					Low	Density of stitches - tearing of paper	Thin sheets used in tensile structure

Table 2.16 Hybrid joints

Nr.	Joining Technique	Transferability of Forces				Tech-nology	Critical Area / Aspect	Recommended Application
		P	T	S	M			
1	3d-printed Node					Low	Limited strength	Small-size structures built indoors
2	Direct Form-locking Node					Low	Contact areas, durability (softening)	Limited impact requirements
3	Crossed Tim-ber plates - Node					Mid	Contact areas	Frames with limited torsion
4	Massive wood node					Mid	Bending of beam	Simple orthogo-nal grid-frames
5	Bolted pres-sure plates multi-layered Node					Mid	Contact areas (bolt- paper)	Grid-shell structures
6	Ball node					High	Bending of beam	High-requirement structure
7	Multi-hinged Node					High	Bending of beam	High-requirement structure
8	Gusset node					Mid	Shear and Tor-sion (area of slits)	Low-requirement structure
9	Flower-hinge Node					Mid	Bending of beam	High-requirement structure
10	Simply hinged connection					Mid	Torsion	Reinforcement diagonals or deployable frame
11	Concrete insert					Low	-	Fundament
12	Wooden insert					Mid	-	Various
13	Fiber rein-forced bolted lap joint				*	Low	Hole	Solid board
14	Bolted clamp					Low	-	Solid board
15	Bolted joint with internal shear plate				*	Mid	Contact between basis material and fixation elements	Solid board
16	Locally rein-forced snap connector joint					Low	Limits of adhesive bond	Tensile structures

17	Multi-layered lap joint with pressure rings and tensile knot				Low	Fall-out of fasteners	Tensile structures
18	Double hinge hook				Low	Failure of pinned sides	Tensile structures

Epilogue for the 'state of the art' and direction of the research following

Following the review process, some joining techniques in the field of beam-structures realized with paper-tubes and also from the analysis of joints in conventional building methods, triggered considerations for the development of this research. Some relevant comments are presented hereby.

Joints from bamboo structures based on dry form-locking, tying and fixing with fasteners (nails) present the advantages of simplicity and practicality, but are not as effective for paper-tubes without the implementation of reinforcements. The reasons for this are related to differences in the matrix of the component, the fiber orientation and length etc. The integration of adhesives would improve the performance of the joints, but in this case the ecological factor comes into consideration. On the other hand, most of the steel intermediate connectors reviewed in realized paper-tube structures are fully functional, especially because they are mostly combined with pre-stressing the tubes. However, such joints fit better in highly durable structures rather than temporary ones. Considering the global aim for building with renewable materials, joints designed with wood-products are observed as a more suitable solution which also allows for building methods that could potentially be realized with far more accessible technologies, comparing to the manufacturing of steel nodes. Thinking about innovation and challenging targets, in combination with the aspect of sustainability, connections formed with thin-shell ecologically friendly composites would be an interesting subject for development.

Literature

Fiction-Factory. Wikkelhouse. Retrieved from https://bouwexpo-tinyhousing.almere.nl/fileadmin/user_upload/Wikkelhouse_BouwEXPO.pdf

Holger Seidlitz, F. K., Nikolas Tsombanis. (2017). Advanced joining technology for the production of highly stressable lightweight structures, with fiber-reinforced plastics and metal. *Technologies for Lightweight Structures*(Special issue: 3rd International MERGE Technologies Conference (IMTC)).

Leijten, A. EN1995-1-1: Section 8 - Connections. Retrieved from https://eurocodes.jrc.ec.europa.eu/doc/WS2008/EN1995_5_Leijten.pdf

McQuaid, M. (2003). *Shigeru Ban*. London N1 9PA: Phaidon Press Limited, Regent's Wharf All Saint's Street.

Octatube. Realizing challenging architecture. Cardboard structures.

TU-Delft. (2008). *Cardboard in Architecture* (Vol. 7): IOS PRESS, Amsterdam, NL.

Figure 3.1 Node composed by laser-cut MDF plates, an introduction to the experimental work developed within this research, designed for 'House 1' (see fig. 3.4).

3 Global structural analysis

Keywords: construction typologies, global structural analysis, boundary conditions, paper tube

The global objective of the studies elaborated in this chapter is to support the experimental approach of this research, with the creation of context for the case studies that are presented in chapter 4 and by discussing alternative structural systems.

The discussion starts with a brief reference to the most prevalent construction typologies for paper structures, in paragraph 3.1.1 'An introduction to paper structures', as these are identified through the review process in chapter 2 'Joining techniques - State of the art'. A variety of building methods is discussed and representative examples of realized structures are provided.

Later the focus on beam-structures composed of paper-tubes is explained, on the grounds argued already in chapter 2 and more specifically paragraphs 2.1.1 'Collection of reference projects and studies', 2.2.1.2 'Framing the review of joining principles and techniques', 2.2.4.1 'Joining techniques for paper-based tubes' and the global observations by the end of the review, as expressed in paragraph 2.4 'Global overview of joining techniques for paper structures'.

Afterwards, in paragraph 3.1.2 'Paper tubes as a construction material', the main component used for the elaboration of studies and experiments is presented, with a focus on the most crucial aspects for this research in relation with the material composition, manufacturing process and material behavior. This is a round profile of a paper-tube with inner diameter 100mm and wall-thickness 10mm.

In continuation, in paragraph 3.1.3 'Construction outlines and typical cases' three typical cases of beam structures: an orthogonal grid, a truss and a dome structure are presented as the context for the potential application of the joining methods that are worked out in chapter 4 'Case studies'. For these cases, global FEM analysis in 'RFEM Dlubal' software is performed, a tool that is commonly used to analyze structural systems composed of beams. The aim is to show the distribution of internal forces in the structural systems and depict the function of the joints. The resulting graphs (N, V_z and M_y) are provided to support the discussion.

By the end of the analysis, an example of a node is selected as the context for the case studies elaborated in chapter 4. This will define the design outlines and the input for the local FEM analysis, for the joining techniques that are examined in detail.

© The Author(s), under exclusive license to Springer Fachmedien Wiesbaden GmbH, part of Springer Nature 2022
E. Kanli, *Experimental Investigations on Joining Techniques for Paper Structures*, Mechanik, Werkstoffe und Konstruktion im Bauwesen 60, https://doi.org/10.1007/978-3-658-34501-3_3

3.1 An introduction to paper structures

3.1.1 Construction typologies

The most prevalent applications regarding the use of paper products for construction purposes are presented hereby. As paper construction is a novel subject, the typologies of structures identified in this case are not the same as the ones presented in common construction methods, but instead regard small-scale structures and are influenced by the limits of paper products in use. A conceptual categorization of these is made for the purpose of clarifying the context in which this research is performed and explain the decision-making process. This way, the main construction typologies identified, based on the review performed in chapter 2 and the selection of reference projects and studies presented there (paragraph 2.1.1), are the following: plate, shell and beam structures. Further sub-categories and hybrid typologies include thin walled origami, tensile and lamella structures (see table 3.1).

The composition of the structural components and the global structural system, in the different construction types, play a significant role for the development and effectiveness of the joints. Below some representative alternatives are discussed, with the aim to highlight the pros and cons per case.

The categories of plate and shell structures mainly regard loadbearing multi-layered sandwich panels. The layering of these panels can be developed according to the specific function of them (part of outer or inner wall, roof etc.) and the requirements set for the construction. The stiffness of the plates vertically (z) and perpendicularly (y) to the main surface is very important (see table 3.1). The strategic integration of different paper products in the multi-layered panel could increase structural stability. For instance, paperboard is often selected as a reinforcement that functions for both cases mentioned above and honeycomb board particularly for forces acting perpendicularly to the surface, such as wind-load. These construction typologies are ideal for prefabrication, especially due to the fact that the lamination process requires specific conditions and machinery to achieve optimal results. Hence, modular or fully pre-fabricated structures are some of the existing design and assembly concepts, with further potential for relocation of the construction. The 'Wikkelhouse' project (Fiction-Factory) is a representative example of this kind, characterized by a hybrid construction typology.

Within the typology of shell structures certain forms with closed geometries, such as domes and arches, can enhance structural stability, by distributing the forces along the structural members more evenly, keeping these under compression and directing the wind flows aside from the facades of the building-skin, through aerodynamic shapes.

Therefore, such structural concepts are popular for temporary installations and demonstrators of innovative building technologies. An interesting example of this kind is a cellular cavity structure with hollowed panels formed with paper-board (Yuan 2018). An aspect that can be challenging is the fabrication, when there is a high number of panels with different geometrical characteristics. In principle, when the geometry is optimized for division in only a few groups of panels with identical shape, high efficiency for the manufacturing process is possible.

As discussed in chapter 2 and specifically in paragraphs 2.1.1, 2.2.1.2 and 2.4.2 (epilogue), the typology of skeletal structures, realized with paper-tubes is an alternative solution for providing a load-bearing frame for lightweight types of cladding. For the structural system, most commonly, an orthogonal grid may be preferred for space efficiency. However, similarly to previous typologies, arch and dome structures are often preferred due to structural advantages, especially for installations with greater size, such as temporary theaters, event venues etc. For the main structural component, a variety of options, such as profiles or multi-layered composite beams, as discussed in chapter 2, are some of the possibilities. The numerous examples of skeletal structures composed of paper-tubes witness the popularity of this product for structural applications. Furthermore, the relative flexibility of this component, seems to make it attractive even for the generation of multi-layered grid-shell structures (Japan Pavilion (Miyake 2009), see construction detail in fig. 2.28, paragraph 2.2.4.1).

A selection of reference projects and prototypes that are mostly already addressed in chapter 2 are reviewed below, to highlight the potential of the respective construction typologies, this time focusing on the global structure and building method.

Several attempts to build tiny living spaces with boards have been made, both by companies, but also the research community. Project 'Pappeder 26'(Voigt, 2007), realized in 1972 is an early example of this kind. Modern designs often combine paper products with common building materials, such as timber (structural reinforcement) and aluminum (rain protection). Two representative examples in this spirit are a school building in UK (Cottrell-Vermeulen-architecture, 2001) and the Wikkelhouse (Fiction-Factory), shown in figures 3.2 and 3.3 respectively. The second one includes full prefabrication of segments, optimized for easy assembly. Within the research community, examples such as the final prototype presented in the study titled 'Von der Faser zum Haus' (Schütz, 2017) and the prototype ‚House of Cards' (Latka, 2017), are efforts to develop further the application of paper products in modular housing. The process of manufacturing multi-layered panels is still under development. Moreover, interesting studies on the structural performance and the effects of humidity of boards can be found in the field of transporting goods in cardboard boxes.

Paper and thin boards are ideal products for creating deployable thin-walled – origami – structures, as they are easy to crease (Jackson, 2011), that are often useful in humanitarian projects and also for teaching purposes. The foldable pattern can create special stability effects. Nonetheless, as the creased edges are susceptible to fractures, the durability of origami shelters is limited. Still, reinforcement of the edges could increase the lifespan.

Table 3.1 Prevalent construction typologies for building with paper-based products. This table is based on the findings from the review process (chapter 2). The typologies introduced are addressed here briefly. Structures built with profiles are clustered under the categories of beam structures composed of paper-tubes and lamella gridshells composed of composite thin profiles. Structures built with multi-layered composite surfaces are clustered as following: Shell-structures with organic form and plate structures with orthogonal grid composed of thick composite panels, origami structures shaped with thin composite boards and tensile structures made with thin composite sheets. For each case the areas of the primary joints are indicated on the global models with red colour and principle joining techniques are presented.

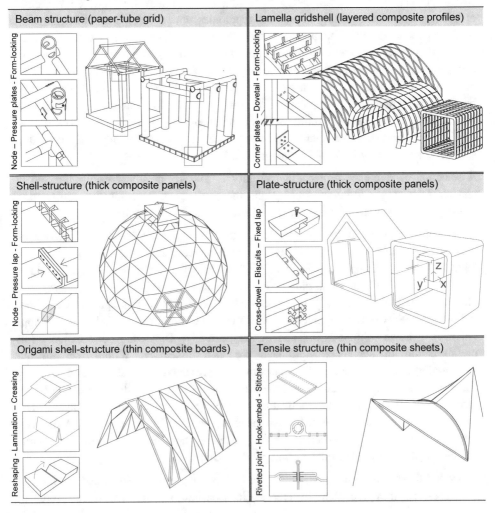

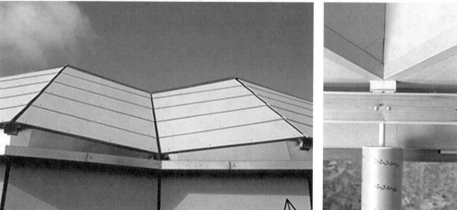

Figure 3.2 Westborough Cardboard Building, 2001, Cottrell & Vermeulen architecture (author: Peter Grant photography). This design uses cardboard components especially for the cladding, immitating an origami structure. As seen at the bottom-right image, paper-tubes are used as columns and are connected with the timber beams with steel rods. As explained in paragraph 3.1.2, paper-tubes present their highest performance under axial compression, whereas bending is one of the problems, a reason that explains the selection of timber beams in this project instead.

Figure 3.3 'Wikkelhouse', a project developed by 'Fiction Factory', a company placed in Amsterdam (NL). The construction combines the principles of shell and plate structure. The connections between the modules are solved with a combination of form-locking and pressure. (Top) installed module (Rotterdam-port, NL – image captured in September 2019), (bottom) production method: a rotating mold, used to wrap around and laminate multiple layers of corrugated paper. This reference project is presented in chapter 2.2.3 at first, with a focus on the assembly process (fig. 2.21 - 2.23).

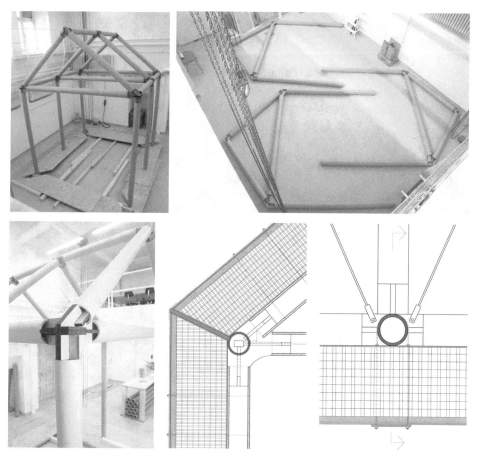

Figure 3.4 'House 1', Full-scale prototype and construction details of beam-structure, built in spring of 2017. The development of the joints (lightweight timber nodes based on form-locking) and assembly method were an experiment, as an introduction to this research.

In the typology of skeletal structures composed of paper-tubes, the most recent and modern design of a space-frame for long-term use is shown in figure 3.5 (bottom). It is the most high-tech structure of this kind that presents a new generation of nodes. In principle, the combination of steel nodes with bars that prestress the paper-tubes make the pre-assembly of the construction in parts more feasible, as shown in figure 3.5.

Sticking to the approach of using renewable materials and simple manufacturing methods, to build temporary structures, 'House 1' is the first experimental case study developed in the first four months of this research. The main idea for the assembly of the structure is to prefabricate all sections and then connect them on-site, with the in-between beams. The connections were designed according to this goal (see appx. 6.1).

Useful references for the design of the joints are presented in paragraph 2.2.4.1 'joining techniques for paper-based tubes'. In parallel with this case study several issues that are important for the structural performance were studied. The performance of the paper tubes for axial compression and bending are two of them. Principle structural analysis of the joints was also attempted. More information regarding this process is presented in the relevant conference papers (Kanli, 2019) and (Nihat Kiziltoprak, 2019).

Figure 3.5 In the beam structures presented above, the combination of steel nodes together with prestressing the paper-tubes, allow for pre-assembling the construction in units. (Top) 'Paper tube dome' (2004) – architect: S. Ban – and (bottom) 'Roof structure at Ring Pass, Delft' (NL), author: 'Octatube', sources: (Eekhout, 2010), (Octatube, 2010).

The focus on beam structures

This paragraph is the bridging point to the next phase of the research, in which the ultimate focus lies on paper-tubes and joints for beam-structures. This is the concept expressed also in the first experimental study presented in figure 3.4. In the previous paragraph alternatives on construction typologies are discussed that include a wide range of paper-based products and components. A set of different issues influenced the reasoning behind the selection of the focus on beam structures.

For instance, despite the ultimate focus of this research on joints, gaining understanding about the behavior of the structural components is essential to make progress on this subject. This way, aspects such as the load bearing capacity of the paper-based components themselves, the durability in different environmental conditions and the range of applicability, based on these factors and also in combination with the design aspect, are important input for this research. In this spirit, certain advantages of paper-tubes comparing to other paper-based products for the selected application were considered, as explained below. The same arguments are also discussed in paragraph 2.1.1 and are mentioned here again for the benefit of keeping the decision-making process visible.

The precedents of realized projects presented in the tables 2.1 and 2.2, in paragraph 2.1.1 'collection of reference projects and studies', indicate the strong influence of paper-tubes in the field of building with paper. Paper-tubes are easily accessible products that require minimum post-processing. In other words, a paper-tube is the only paper-product that is produced in scale 1:1 (for building purposes). Hence, it is treated as a finished structural element, a fact that makes it very attractive for designers. For the moment, it is the paper product with the most satisfactory structural performance for such an application, even though, comparing to common construction materials, it is still not very strong. Specific information about this aspect is provided in the next paragraph (3.1.2). Next to this, its solid wall creates good conditions for humidity resistance and prevention of delamination, comparing to the lightweight corrugated- and honeycomb- boards. This point is crucial for the development of a relatively durable primary construction.

3.1.2 Paper tubes as a construction material

A variety of construction principles for temporary shelters and also more durable structures are discussed in paragraphs 2.1 'groundwork' and 3.1.1 'an introduction to paper structures'. Here important information regarding the manufacturing and structural performance of paper-tubes is presented that is particularly useful for the research process following.

Paper tubes are manufactured with winding, as shown in figure 3.6. An advantage of this process is the easy adjustment of the wall-thickness and diameter of the core. Most commonly the winding process is performed with an angle. The great advantage of this process is that the tubes are cut in the desired size at the end of the production. However, the winding pattern has drawbacks on the structural performance, as it limits the strength of the tubes for bending and buckling. Moreover, the tubes may be susceptible to the effects of torsion, in the direction of winding. Still, parallel winding (figure 3.6, right) is not popular and is used for tubes with small size.

Further products of paper-tubes are currently being researched, aiming either for resolving the issues stated above or for creating more lightweight beams by integrating hollow boards in the design of the wall.

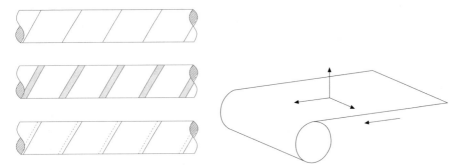

Figure 3.6 (left) Details regarding the production of tubes[1] - lamination principles from the bottom to the top: overlapping, with distance, edge to edge (0 seam)-, (right) Parallel winded tube

The component selected for this research

A wide spectrum of products that present different qualities is available by the industry, depending on the base paper material and the adhesive used for the lamination. For this research one specific component is selected, to create suitable conditions for comparing different joining methods. The dimensions of the section are the following: outer diameter (120mm), wall thickness (10mm). The mechanical properties assumed are presented in table 3.2. The value for stress max. is based on the results of structural tests that were performed by the institute of ISM+D (Kiziltoprak, 2019). Extensive information regarding the testing conditions are presented in the conference paper (Kanli, 2019). Due to the relatively small diameter of the tubes it is expected that the axial strength defines mainly the performance. These results lie in the same range with the findings presented in the book 'Shigeru Ban', where some details about structural

[1] Source: Blechschmidt, J. Hrsg.: Papierverarbeitungstechnik. Carl Hanser Fachbuchverlag, München, 2013

tests for the performance of paper-tubes, made during the development of projects, are provided (McQuaid, 2003, p. 74). For the aspect of material imperfections, a difference of ±5mm for the inner diameter of the tube is possible. However, no such big deviations were observed in the material purchased.

Table 3.2 Physical and mechanical properties for the paper tube used in the experiments presented in chapter 4.

0	Property	Value	Unit
1	σ_{yield}[2]	9	MPa
2	Young's modulus	728	MPa
3	Poisson's ratio[3]	0,2	-
4	Specific weight[4]	691	Kg/m^3

3.1.3 Construction outlines and typical cases

In this paragraph three typical cases of structural systems, built with paper-tubes, are presented that are further examined in paragraph 3.2 'global FEM analysis', where the objectives of the analysis are described in detail:

- an orthogonal grid (similar to the realized projects 'Paper Log House(s)', 'Temporary school(s) in China', 'Shelter in Haiti' (Jacobson, 2014)),

- a truss-structure (with reference to the projects 'paper arch' (McQuaid, 2003) and 'temporary studio in Paris'(Miyake, 2009))

- a dome structure (with reference to the project 'Cardboard dome' (TU-Delft, 2008))

The projects referenced above are reviewed within chapter 2, in paragraph 2.2.4.1 'Joining techniques for paper-based tubes'.

These typical cases create some interesting alterations in the design of the joints, as the same basic matrix is transformed in three forms starting with a linear (fig. 3.7), to a rotational (fig. 3.8) and from there a translational (fig. 3.9). The whole point of examining them is to become acquainted with the functionality of the joints in the different cases, as described in paragraph 3.2.1 'guidelines for the FEM analysis' and define the most important aspects for the effectiveness of the joints, in relation with the selected structural component. Gaining this understanding is very useful, prior to the development of case studies.

---

[2] This value has been calculated based on the results from a material test for compression.
[3] (Nihat Kiziltoprak, 2019)
[4] Details about calculating the density can be seen in (Iggesund-paperboard, p. 112)

A set of priorities is formed regarding the dimensioning of the beam structures and also the function of the joints, based on the current knowledge about the main component and its structural behavior. In principle, the idea is to eliminate possible stability issues by dimensioning the structures in a sensible way, despite the lack of rules or standards. For example, the span of the beams is at maximum 2 m. Next to this, in all three typical cases, bracing elements are used for tension. Further details are provided in paragraph 3.2, where the conditions for the global FEM analysis are described.

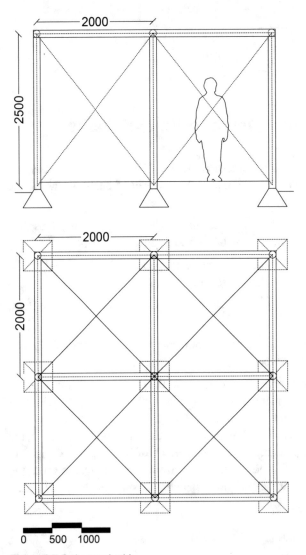

Figure 3.7 Orthogonal grid

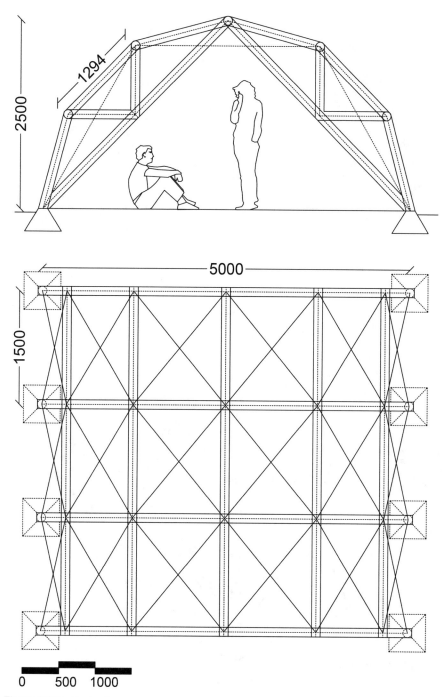

Figure 3.8 Truss structure

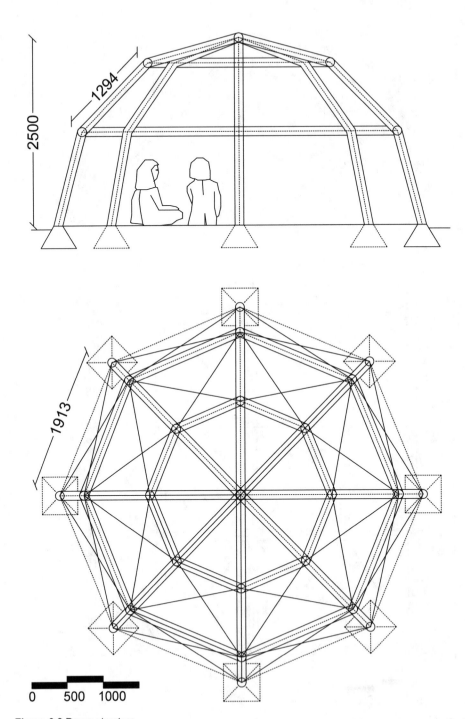

Figure 3.9 Dome structure

3.2 Global FEM analysis

The aim of the analysis is to examine the distribution of loads in the typical cases and select a scenario that will define the boundary conditions for the case studies, meaning the design outlines and the input for the local FEM analysis. The software used for this purpose is 'RFEM", by Dlubal'. The application of loads is presented individually for each typical case. The method and steps taken to build-up the models and define the boundary conditions are explained in paragraph 3.2.1. In principle, the most important aspects are related to the definition of materials, structural members, load cases and calculation methods. The three typical cases are developed within common grounds, with respect to, both, the composition of the basic structural system and the relevant input required particularly for the materials and the calculations. More specifically, the mechanical properties applied for the component of the paper tubes are these stated in paragraph 3.1.2, table 3.2. The characterization of the joints is also very important for the distribution of internal forces. For the global analysis the joints are simply identified as fixed or hinged, with no further specifications implemented as input. For the stability of the structural systems, bracing elements are applied, for tension, in order to decrease the displacements. This decision is related to the performance of the rotational stiffness of the joints. As mentioned in paragraph 2.3.4 the characterization of joints for steel is advanced, for example comparing to timber, due to the predictable material behavior that has allowed in depth understanding of the mechanics. Interesting input on this subject is provided in Eurocode 3-1-8, for steel structures (1993-1-8, 2005).

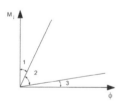

Figure 3.10 EN 1993-1-8:2005, p. 57

There, joints are classified by their stiffness, in three zones: 1 "rigid", 2 "semi-rigid" and 3 "nominally pinned" (flexible) and the requirements for the realization of this classification are defined. The graph in fig. 3.10 that shows the relation between the moment applied on the joint and the resulting rotation, is used as a reference. For example, based on draft testing results and simplified calculations, it is estimated that for typical case 1 (orthogonal grid, 3.2.2) the joints could be classified as rigid when combined with bracing elements, without further necessity to provide a fixed value for the rotational stiffness (see table 3.4, φ_χ). In paragraph 4.3 interesting comments on this aspect are made, based on the results from structural testing performed within the case studies. All in all, this process is expected to create better understanding over the structural issues and to create context for the case studies (chapter 4). On the other hand, any changes in the dimensioning of a structural system could have significant impact on the resulting loads. Thus, the idea is to obtain a spectrum of values. For the present studies it is considered that the primary structure is protected from alternating weather conditions. This condition is necessary to assume a durable structure.

3.2.1 Guidelines for the FEM analysis

Build-up of models[5]

Below the main steps followed to set-up the models and the parameters taken into consideration are described.

Step 1: Define the nodes required to create the beams for the primary structure.
Step 2: Define the materials to be applied.
Step 3: Define the cross-sections needed to create the structural members (in this case the dimensions of the paper-tube component presented in 3.1.2 apply).
Step 4: Create 'Member Hinges' to be applied at the start and end of each beam.
Step 5: Create Members
Step 5.1: Create beams: paper-tubes
Step 5.2: Create members for bracing: Tension
Step 6: Set Nodal supports as foundation
Step 7: Create Load cases
Step 8: Create Load Combinations
Step 9: Calculate

Hereby, insightful information regarding the input provided to the software, within the steps presented above, is provided: In step 1, the dimensions of the skeleton are already defined by the position of the nodes. As the typical cases have different geometries, the length of the beams is not equal. The idea is to design structures with different geometry that provide interior spaces of similar size. In continuation, in chapter 4, the potential application of the case studies in different scenarios, by adjusting the design, is an aspect considered as well. In step 2, the input on mechanical properties of beams is set as shown in table 3.3. In step 3, by definition, the cross section of the beam is considered as a perfect element, a condition that deviates from reality. Imperfections are implemented when this is considered critical (see typical case 1, paragraph 3.2.2 'Orthogonal grid').

Table 3.3 The input on material properties used for the global structural analysis in 'RFEM' software

Material	Modulus of elasticity E (kN/m^2)	Specific weight γ (kN/m^3)	Poisson's Ratio ν	Partial safety factor γ_M
Paper tube	728000 (E_{axial})	6,91	0.2	1.2
Steel S235	210000000	76.98	0.3	1

Table 3.4 Definition of the behavior of 'Member Hinges' in 'RFEM' software

Reference system	Axial/ Shear Release or Spring (N/mm)			Moment Release or Spring (Nm/°)		
	u_x	u_y	u_z	φ_x	φ_y	φ_z
Local x,y,z	No	No	No	No	Yes	Yes

[5] Many terms in this paragraph are named after the respective commands or functions in the software 'RFEM, by Dlubal'.

In step 4, the member hinges are supposed to define how the beams are connected with each other. Commonly, these are assumed either as fixed or hinged. When a joint is assumed to be fixed, this means that its rotational stiffness is infinite, a condition that would minimize the stresses and deformations in the beams. On the other hand, when the joint is hinged it is free to move or rotate in certain directions, based on the conditions assumed. Basically, the total deformation is minimized when all joints are fixed and maximized when the majority of joints are simply hinged. On the other hand, the internal forces that are concentrated on the node follow the opposite pattern. Both scenarios are examined and the range of values between the results is presented for each typical case. The implementation of tensile elements for stiffening minimizes the gap between the results extracted based on these two different conditions. In steps 8 and 9 two load cases are considered, dead load and wind (ASCE 7-16). The magnitude of the loads and the distribution along the members are presented in each typical case. Each load case is calculated with 'geometrical linear analysis', whereas the load combinations with 'second order analysis'. For the load combination, both load cases are superimposed with a factor equal to 1.0. In addition, the 'RSBUCK' function carries out a 'stability analysis' that detects whether there is failure of the system due to buckling. The 'critical factor f' is used as an indicator for the stability of the system. Further differentiations per case can be seen in the respective paragraphs.

Further considerations

Material imperfections: Member Nonlinearities are excluded, due to limited input. For example, for the paper tubes, the section is not homogenous along its length, as a result of the manufacturing process (see 3.1.2). Next to this, further research is necessary to define the properties in three directions (axially, perpendicularly and radially) and have complete information for the structural performance of the component.

Load cases: In the current studies, hypothetical load cases are applied, aiming to set a realistic benchmark. According to standards, for the superposition of all load cases, unfavorable conditions should be considered, by implementing, for example, a scenario in which the dead load is multiplied by a factor equal to 1.35 and the wind load by 1.5 (CEN, 2003) etc. Further studies on profiles with different properties (diameter, wall thickness etc.) are required, to examine if such standards could be satisfied.

Stiffening Methods: In this study, bracing elements for tension are used. To develop the structural analysis further, the possibility to integrate plates, as walls that act as stiffening elements for the frame, has been considered. Even lightweight timber or particle boards could create a positive effect. Precise dimensioning of the construction, in full detail, is required prior to building such a model for FEM analysis. Other stiffening methods include either densification of the structural grid, or integration of a sub-grid or stiffening diagonals.

3.2.2 Orthogonal grid beam-structure

The context for the case studies will be selected from this orthogonal grid, based on the following grounds. It is the most challenging case, with the highest potential for wide application, due to space efficiency and potential expansion of the structure to a second level. Thus, higher loads are applied, comparing to the other typical cases, with the aim to create a higher benchmark for the performance of the joints (chapter 4.2).

Load cases

The distribution of loads on the members can be seen in table 3.5 and fig. 3.11. On the roof, simplified uniform distribution of loads on the beams is applied. In a different way, on the façade, where the wind load acts, the distribution of loads is made with the use of trapezoids, as the higher part is expected to receive the majority of the loads. Imperfections are applied on the members that are primarily subjected to compression, so all columns, as it can be seen in fig. 3.11 (right). The scenario applied regards deformation of the columns in the same direction as the wind-load.

Table 3.5 Distribution of 'Member Loads' in typical case 1 (orthogonal grid)

Load case	Location	Area (m^2)	Total Load (N)	Total length of beams (mm)	Load (N/m^2)
Weight	Roof - Perimeter	13.28	4800	16000	~ 753
	Roof - Center		4800	8000	
Wind	Façade- Columns	10	2250	7500	345
	Façade - Beams		1200	4000	

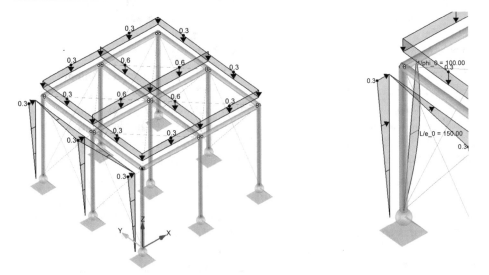

Figure 3.11 (left) Distribution of loads, (right) Implementation of imperfections

Structure and connections

On the level of the eave, in every junction between the beams, only the end of the column is assumed as fixed. So, all beams are hinged at both ends. The nodal supports (fundaments) are always fixed points.

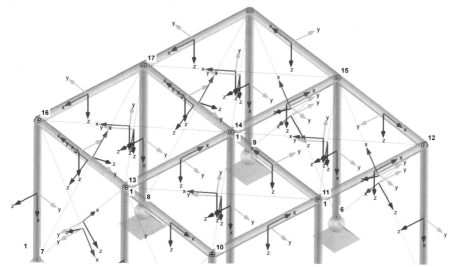

Figure 3.12 Structural system and local axes of co-ordinates

The load case expressed in z-direction has significant impact on the cross that is shaped in the middle of the structure and its fixations. It needs to be transferred efficiently to the compression members, in order to avoid bending of the paper tubes. This unfavourable case is represented by the scenario of hinged joints. Next to this, the wind causes significant displacements in the x direction.

The tensile elements minimize the displacements, improving the transferability of forces within the structural system and stabilizing the level of the eave. When no tensile elements are applied, the specifics regarding the performances of each beam and joint are much more important (mechanical behaviour). As a consequence, in that scenario, the idealization of fixed joints is essential to achieve stability and prevent buckling of the structure.

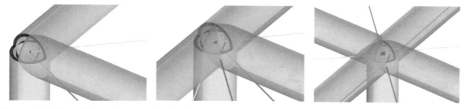

Figure 3.13 The different cases of nodes in the system (left) corner, (middle) middle, (right) center

Results

Hereby, the results of the calculations made according to the conditions described in the previous paragraphs are presented.

As shown in fig. 3.14 the highest deformations occur in axes z and x, due to bending of the beams. As stated in table 3.6, when all joints are hinged the maximum deformation is 50,7cm, with this value being appointed at the center of member 9 (fig. 3.20, right). The same beam presents the highest rotation in axes x and y. When all joints are fixed, then the maximum deformation is significantly decreased and is about 26,9cm. The high deformations are also a result of applying imperfections to the model.

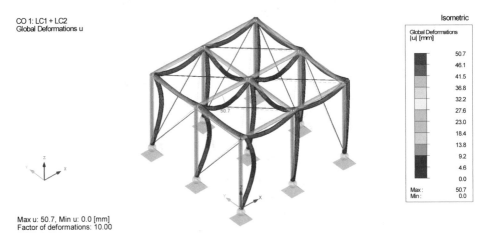

CO 1: LC1 + LC2
Global Deformations u

Isometric

Global Deformations
|u| [mm]

50.7
46.1
41.5
36.8
32.2
27.6
23.0
18.4
13.8
9.2
4.6
0.0

Max :   50.7
Min :   0.0

Max u: 50.7, Min u: 0.0 [mm]
Factor of deformations: 10.00

Figure 3.14 Visualization of global deformations (factor of deformation 10) when the joints are set as hinged.

The critical factors calculated by the add-on module 'RFSTABILITY', as these are presented in table 3.6, indicate that the structure does not buckle and in theory the amount of loads applied could be increased by two times approximately. However, as in fact the paper tube is not a linear elastic component, but instead an orthotropic one, with certain imperfections, one to one comprehension of the results is not possible. Hence, the fact that the critical factor is higher than 1 is perceived as a relatively positive result, but it is, in principle, desired to obtain results that stand as far as possible from the limits of the structure and the material itself.

Table 3.6 'RFSTABILITY', a method to evaluate the stability of the structure through the critical factor f and maximum deformation, as calculated in RFEM software

Value	Fixed	Hinged
f	2,941	2,181
$U_{t.max}$ (mm)	26,9	50,7

Figure 3.15 Observing the deformations around the central node when the joints are hinged.

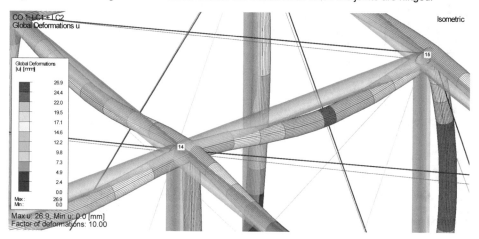

Figure 3.16 Observing the deformations around the central node when the joints are fixed.

In paragraph 3.2.1 (table 3.4) the definition of the joints is discussed and particularly the rotational stiffness. Hereby, a representative example regarding this issue is demonstrated. Node 14 is a critical area for the stability of the construction. Figures 3.15 and 3.16 are used to visualize the impact of the joints (hinged and fixed) on the distribution of loads and the resulting deformations. The maximum displacement in x-direction and y-secondarily- is higher. The column below node 14 presents higher bending in x-direction. Thereby, the stiffness of the beam itself suddenly is by far more significant. On the contrary, in the case of fixed joints, the nodes transfer the forces effectively and receive the highest moments, minimizing thus the impact on the structural members. The same effect is demonstrated for the entire structure in figure 3.19.

Distribution of Internal forces

Here the output from the calculations is provided. The exported 3d-graphs are the results from the calculation in which the nodes are set as hinged, that is the most unfavorable condition. These diagrams provide useful information for the functionality of the structural members and nodes. Next to this, the boundary conditions for each joint within the current system become known.

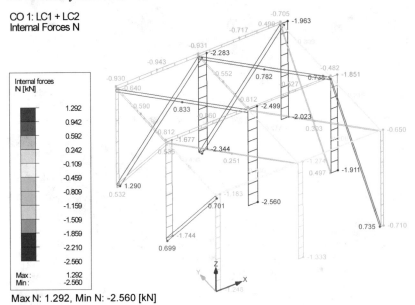

Max N: 1.292, Min N: -2.560 [kN]

Figure 3.17 Distribution of Normal forces in the structural system (model with hinged joints). The signs (+ or -) are defined according to the local co-ordinate systems that apply for each individual structural member (fig. 3.12).

In the diagram of Normal forces (fig. 3.17), the color map on the left presents the range of forces carried by the members. According to this diagram, there is a concentration of normal forces on compression members 13, 19, 18 and 17 but also the beams 6 and 9 primarily and 5, 7, 12 and 10 secondarily. The visible cables are the only ones that contribute to the stiffening of the structure. Following these, the flow of normal forces in the structure is better comprehended. The numerical values indicate the significance of the presence of these elements for the stability of the structure. These results indicate that the nodes need to resist significant shear forces. Additionally, to achieve proper function of the tensile elements, a strong fixation of those on the joints is important. This can become a rather challenging aspect, as high concentration of stresses will appear at the points of fixation, considering a force up to 1kN approximately that is pulling the joint.

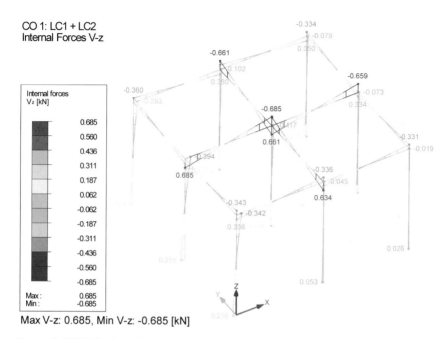

Figure 3.18 Distribution of lateral forces in the structural system (hinged joints)

The diagram of lateral forces, presented in fig. 3.18, shows that the impact in z-direction is significant, especially for the nodes that form a cross in the middle of the structure, on the level of the eave. This indicates that the joints need to be bending-stiff in z-direction to withstand the impact from shear forces.

The 3d diagrams in fig. 3.19 (top and bottom) show the moments M_y in both cases of hinged and fixed joints respectively. These visuals create a good impression regarding the distribution of the moments in the system, in these two different cases, as described beforehand. So, in the first case, in which the joints are hinged, the moments are maximized in the middle of the beams, whereas in the second on the joints. For the global analysis, in principle, the assumption of hinged joints is preferred, as an unfavorable concept for the stability of the structure overall. The performance of any kind of joint designed for this structural system lies between these two cases.

Overall, in principle, the minimization of lateral forces or moments that could cause bending of the tubes is important, due to the fact that, as structural experiments have shown, a paper-tube is much stronger under axial compression than bending. Keeping in mind the hygroscopic behavior of paper that would lead to softening of the component through time, this position becomes even more important, even though these effects are not studied within this research.

CO 1: LC1 + LC2
Internal Forces M-y

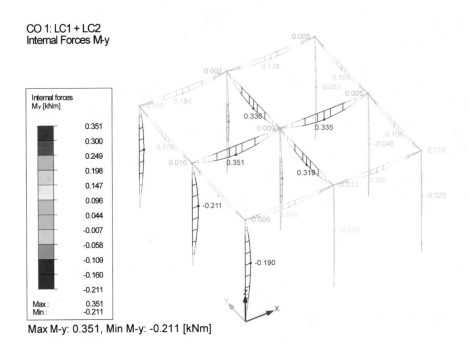

Max M-y: 0.351, Min M-y: -0.211 [kNm]

CO 1: LC1 + LC2
Internal Forces M-y

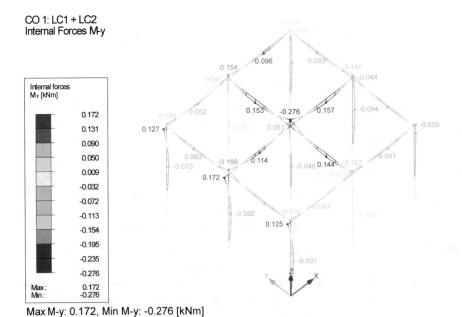

Max M-y: 0.172, Min M-y: -0.276 [kNm]

Figure 3.19 Distribution of moment M_y in the system with hinged (top) and fixed (bottom) joints

Priorities and issues that require further development

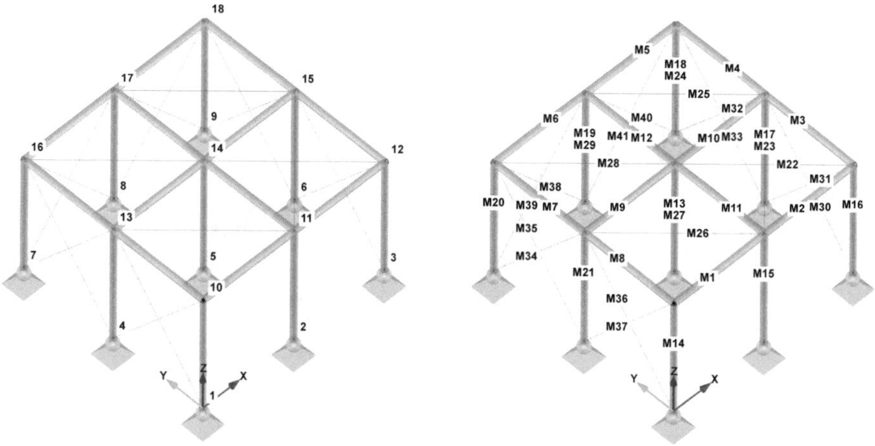

Figure 3.20 Labelling of 'Nodes' (left) and 'Members' (right) in RFEM, to assist the observations

An important issue for the joints is to prevent bending by absorbing the moments in y-direction and the lateral forces in z-direction. Effective distribution of the normal forces through the joints is essential for the stability of the structure. The geometry of the tube creates difficulties in achieving this condition (for example increased contact surface between the structural elements is important). For details regarding the context selected for the case studies see paragraph 3.3.2.

Critical Nodes and Members: Following the distribution of internal forces in the structural system, the nodes 14 and 13 are the most critical ones and also nodes 17, 15, 11 and 16. Member 13 receives the highest amount of normal forces, and thus presents the highest deformation in z axis due to compression. Still the condition of the members that form the central cross (9-12) is more critical, due to the weakness of paper-tubes when subjected to bending.

Potential optimization: Based on the analysis of the model with hinged joints, strategies for the minimization of the maximum deformations presented on the beams should be considered. These could include:

1. Considering a paper tube with higher bending strength. For example, examining tubes with higher wall-thickness or diameter or both. Selection of different base material could also make a difference.

2. Re-dimensioning the structure by decreasing, for example, the spanned distances to less than 2 m that is the current measurement.

3. Redesigning the structure with a pitched roof, to distribute the forces better between the central and the side columns.

4. Examine the possibility of using plates for stiffening.

3.2.3 Truss structure

Structure and load cases

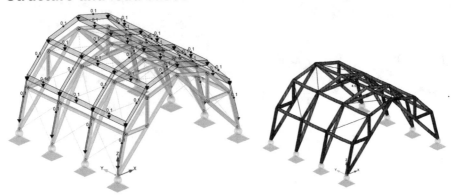

Figure 3.21 Vertical loads applied (left) and resulting deformations (right) – hinged joints assumed.

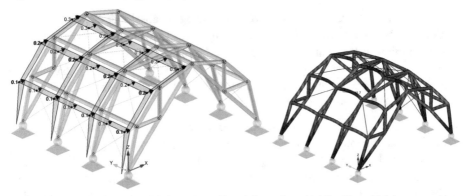

Figure 3.22 Wind loads applied (left) and resulting deformations (right) – hinged joints assumed.

The same steps, as in paragraph 3.2.2 'orthogonal grid' are followed, with the aim to discuss the distribution of loads on the structural system. To span a distance of 5m (main elevation), the integration of stiffening diagonals is necessary. The dead load creates the highest deformations on the peak of the arch, at the horizontal stiffening beams within the arches and at the beams that connect the consecutive arches. The wind-load affects mainly the beams between the arches.

Table 3.7 Distribution of 'Member Loads' in typical case 2 (truss-structure)

Load case	Area (m²)	Load (N)	Total length of beams (mm)	Load magnitude (N/mm)	Load (N/m²)
Dead load	35	5356	53556	0.1	~ 153
Wind load	17.5	3735.2	29028	0.1-0.3	~ 213

Results

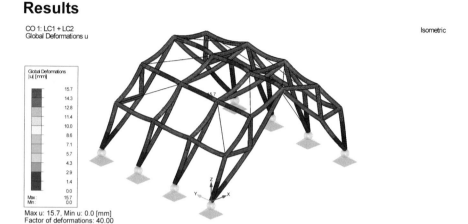

CO 1: LC1 + LC2
Global Deformations u

Isometric

Global Deformations
|u| [mm]

15.7
14.3
12.8
11.4
10.0
8.6
7.1
5.7
4.3
2.9
1.4
0.0

Max: 15.7
Min: 0.0

Max u: 15.7, Min u: 0.0 [mm]
Factor of deformations: 40.00

Figure 3.23 Global deformations in the structural system with implementation of hinged joints.

Overall the diagonals on the elevations carry the greatest normal forces, whereas the arches the lateral forces and moments. The most critical areas are identified on the upper part of the arch. As it can be observed in the diagrams of internal forces in fig. 3.25, in that area, the normal and the lateral forces are maximized and the beams that connect the consecutive arches present higher bending. Still, the maximum total deformation lies within an acceptable range (table 3.8). The small deviation between the $u_{t,max}$ for fixed or hinged joints indicates the benefits of the closed geometry for the overall stability. Next to this, the difference in the critical factor f is significant, as the geometry of the structural system enhances the role of the joints for the stiffening of the whole structure. This way, the stiffness of the joints could, in theory, be used more effectively to improve stability. However, the integration of stiffening diagonals leads to more complex configurations for the joints (see fig. 24) that could challenge the design process.

Table 3.8 'RFSTABILITY', a method to evaluate the stability of the structure through the critical factor f and maximum deformation, as calculated in RFEM software

Value	Fixed	Hinged
f	5,887	1,126
$U_{t,max}$ (mm)	10,9	15,7

Figure 3.24 The different cases of nodes in the system: (left and center) middle joints (with and without diagonals), (right) side joint (with diagonal).

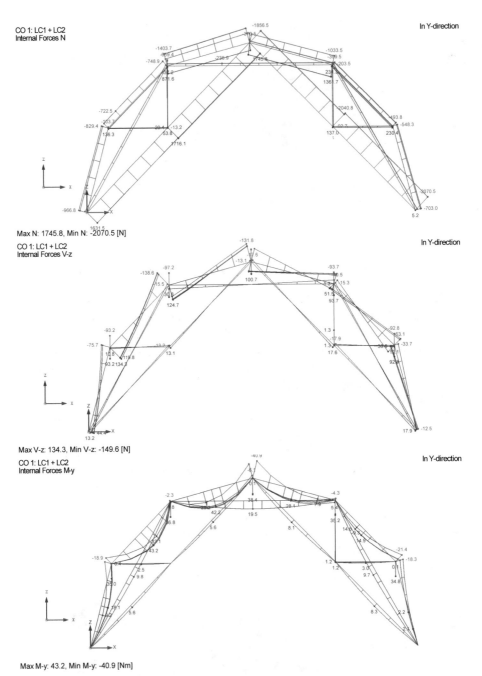

Figure 3.25 Elevation diagrams that show the distribution of normal (N) and lateral (V$_z$) forces and moments (M$_y$) within the truss structure examined. Structural members located further behind the front elevation are also visible, following the structure presented in fig. 3.23.

3.2.4 Dome beam structure

Structure and load cases

In principle, from the three typical cases, this is the most stable geometry. The highest deformations caused by the dead load are observed on the lower horizontal beams (fig. 3.27). The areas mostly affected by the wind-load are these that receive it perpendicularly to the axes of the beams and are further from the supports. The application of the wind-load on the opposite side of the entrance is selected as an unfavorable condition, in order to examine the stability of this opening in the structure.

Figure 3.26 The structural system composed in RFEM

Table 3.9 Distribution of 'Member Loads' in typical case 3 (dome-beam-structure)

Load case	Area (m²)	Load (N)	Total length of beams (mm)	Load magnitude (N/mm)	Load (N/m²)
Dead load	36	10776	67272	0.2	~ 299
Wind load	18	4663	37518	0.1-0.3	~ 259

Results

The load combination in both cases of hinged and fixed joints, leads to a total deformation (U_t) that lies within a low range. The f factor is much higher than in any of the two previous cases. As in typical case 2 (truss-structure), the stiffness of the joints has great impact on the stability. Based on the resulting diagrams presented in fig. 3.28, the normal forces are in principle maximized at the lower part of the structure, beneath the first ring of horizontal beams. Along the first ring of horizontal beams the bending moments are maximized in the middle of each beam and the lateral forces on the sides. Still the values for bending moment and lateral forces lie within a low range, which is positive for the performance of the beams.

Table 3.10 ‚RFSTABILITY', a method to evaluate the stability of the structure through the critical factor f and maximum deformation, as calculated in RFEM software.

Value	Fixed	Hinged
f	9,092	3,366
$U_{t.max}$ (mm)	3,7	10,2

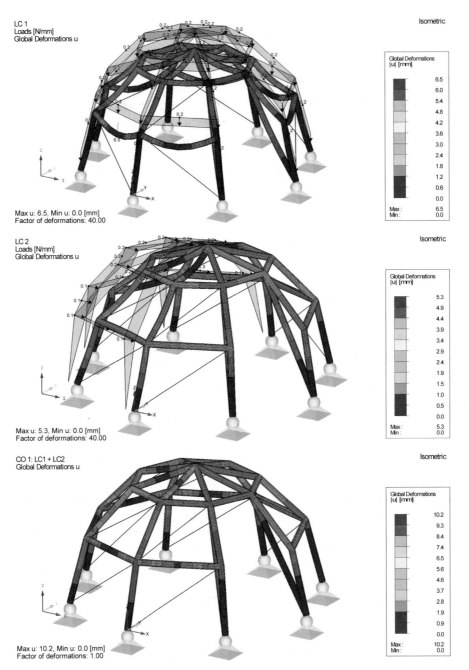

Figure 3.27 Distribution of loads and resulting deformations in the structural system - (top) weight, (center) wind, (bottom) combination of the above – Hinged joints assumed. The factor of deformation and the colour-legend on the right help to comprehend the information presented.

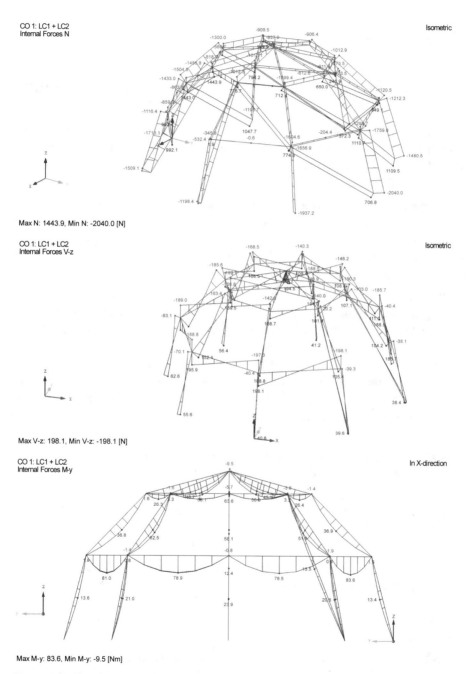

Figure 2.28 2d elevation diagrams that show the distribution of normal (N) and lateral (V z) forces and moments (M y) within the dome structure examined.

3.3 Summary

3.3.1 Observations on the global structural analysis

In this chapter three typical cases of structures with different geometries are examined, as potential environments for the implementation of the joining techniques developed in chapter 4. The aim is to observe the load distribution and draw conclusions for the functionality of the joints.

Each structural system is observed individually. Principle comparisons between the three typical cases help to analyze the desired functionality of the joints. Due to the fundamental geometrical differences, the different spanning distances and structural grids, it is not possible and also not the aim to draw direct comparisons between the values that describe the resulting forces.

At first, some observations about the principle structural systems are addressed. Following the global analysis, based on the stability check 'RSBUCK control' that detects buckling in the structural system, all three typical cases provided positive outcomes (see tables 6, 8, 10). Still, based on these results, in typical case 1 'orthogonal grid' (table 3.6), densification of the grid by decreasing the length of the beams, would probably be necessary, considering that the members were idealized in the simulations. In a different way, the values in the other two cases show that upscaling could be possible, especially for typical case 3 'dome structure' (table 3.8).

The resulting 3d diagrams of internal forces visualize well the different patterns that are also known from basic structural theory, regarding the distribution of loads in the three different geometries represented. To summarize these:

In typical case 1 (orthogonal grid) the compression members take the majority of the normal forces, whereas the resulting lateral forces and moments have a big impact on the beams (maximum deflections). The flat roof leads to critical shear forces on the horizontal plane that challenge the function of the joints.

In typical case 2 (truss structure) the normal forces are more equally distributed between all the beams. The higher distribution of normal forces on the top of the arch -between the two sides - increases the significance of the nodes. As a result of the closed geometrical shape and the short spanning distances the lateral forces and the bending moments are minimized.

In typical case 3 (dome beam structure), the geometry of the structure leads to efficient distribution of the normal forces close to the base. The lateral forces and moments are minimized. The lateral forces are higher on the upper part of the structure and the bending moments are maximized in the longer horizontal beams. Overall, the resulting displacements demonstrated are in total minimized.

Focusing on the functionality of the joints, the global structural analysis comes to underline some important aspects. One of them is to transfer successfully the normal forces and keep the tubes under compression, to minimize other effects. Thereby, from the design perspective, increased contact surface is one of the challenges to overcome, due to the round hollowed geometry of the tubes. Next to this, it is crucial for the joints to absorb the lateral forces and moments that occur on all sides, in order to minimize the effect of bending in the tubes. Moreover, considering the effects of wind-loads that are expected to cause shearing between the structural elements and consequently between the elements of the joints, one more priority is added in the picture.

All in all, in principle, these structures could be feasible under certain limitations. These include aspects related to the load capacity, safety measures (such as tension elements) as well as the durability of the structure (depending also on the climate conditions). Still, due to limited information about the structural performance of the paper tubes in combination with simplifications in the FEM models, the global analysis can only provide an impression for the distribution of loads. At the same time, to proceed further with the aspect of dimensioning structural systems, it is important to evaluate further the structural behavior of different products of paper-tubes, a subject that could be dealt within the fields of civil engineering and mechanics of materials.

3.3.2 Selection of specific context for the case studies

The selection of a specific node for the development of the joining methods within the case studies is mostly related to setting a common context, particularly in regard with the design outlines and for the local FEM analysis performed in chapter 4.

In principle the joining techniques studied in chapter 4 could be adapted to fit for any of the three typical cases. However, at this stage of the research it is important to set a context that would also allow the parallel development of physical experiments (combining prototyping and testing) with computer simulations, to use all means available while proceeding further with the investigations on the design and behaviour of the joints.

Typical case 1, the 'orthogonal grid' is the environment suggested for this purpose. The central connection (node 14) is selected for this purpose. It is the most critical location for the stability of the construction. It also presents the most challenging conditions for the design of the joining method. Furthermore, as mentioned before, typical cases 2 and

3 are also considered within the design of the case studies so that these can easily be transformed to fit in an arch or a dome structure.

In table 3.11, a set of values that describe a load combination for 'node 14', is selected as the boundary conditions for the local FEM analysis performed for selected case studies in chapter 4.2. Hence, this set of values is supposed to function as a hypothetical benchmark.

The use of this input is described in detail under the respective section 'Local FEM Analysis' that is presented for each case study individually. The context for these studies (software, input etc.) is first presented in the start of paragraph 4.2, together with the general guidelines for the elaboration of the case studies. In principle, the idea is to transfer these values in a model where the design of the joint, in this case node 14, is used to perform simulations with the aim to calculate the distribution of stresses within the area of the joint and identify potential failure modes.

Table 3.11 Internal forces applied on 'Node 14' (see labelling in fig. 3.20), in typical case 1 (orthogonal grid) – model with hinged joints.

Member and distance from n.14 (mm)	Forces - F (N)			Moments – M (Nm)		
	N	V_y	V_z	M_t	M_y	M_z
Member 9 (1440)	-902.98	~ 0	-313	~ 0	281	~ 0
Member 10 (560)	-594.5	~ 0	297	~ 0	268	~ 0
Member 11 (1440)	-326.37	~ 0	-281	~ 0	256	~ 0
Member 12 (560)	-615.89	~ 0	298	~ 0	269	~ 0
Member 22 (-)	303	-	-	-	-	-
Member 28 (-)	833	-	-	-	-	-

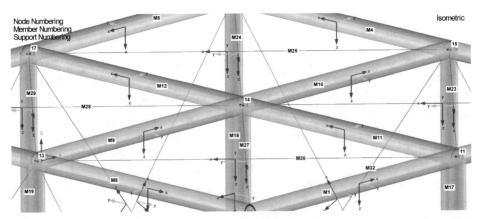

Figure 3.29 Local systems of co-ordinates for each 'Member' that is connected with the critical 'Node 14' (the complete structural system can be seen in fig. 3.12). This image has the aim to support the comprehension of the data presented in table 3.11.

Literature

1993-1-8, E. (2005). Eurocode 3: Design of steel structures - Part 1-8: Design of joints. In.

CEN. (2003). EN 1991-1-3. Eurocode 1 - Actions on structures - Part 1-3: General actions - Snow loads. In.

Cottrell-Vermeulen-architecture. (2001). Westborough Cardboard School Building. Retrieved from http://www.jdg-architectes.com/projet/pavillon-vasarely/?lang=en

Eekhout, N. (2010). Ring Pass. Retrieved from https://www.octatube.nl/en_GB/project-item.html/projectitem/136-ring-pass

Fiction-Factory. Wikkelhouse. Retrieved from https://bouwexpo-tinyhousing.almere.nl/fileadmin/user_upload/Wikkelhouse_BouwEXPO.pdf

Iggesund-paperboard. *Reference Manual. The Paperboard Product.*

Jackson, P. (2011). *Folding techniques for designers from sheet to form.* London EC1V 1LR, UK: Laurence King Publishing Ltd.

Jacobson, H. Z. (2014). *Humanitarian Architecture.* Aspen, Colorado 81611: D.A.P. (Distributed Art Publishers, Inc.).

Kanli, E. (2019). *Case study: Development and evaluation methods for bio-based construction realized with paper-based building materials.* Paper presented at the 3rd Internation Conference on Bio-Based Materials, ICBBM, Belfast.

Kiziltoprak, N. (2019). *Experimental method to determine the bending load capacity of paper tubes.* Retrieved from https://www.ismd.tu-darmstadt.de/forschung_1/material_ismd/bamp____bauen_mit_papier/bamp___bauen_mit_papier.en.jsp

Latka, J. (2017). *House of Cards - design and implementation of a paper house prototype.* Paper presented at the IASS - Interfaces: architecture, engineering, science, Hamburg, Germany.

McQuaid, M. (2003). *Shigeru Ban.* London N1 9PA: Phaidon Press Limited, Regent's Wharf All Saint's Street.

Miyake, R. (2009). *Shigeru Ban Paper in Architecture.* US: Rizzoli International Publications.

Nihat Kiziltoprak, E. K. (2019). *Load capacity testing method for non-conventional nodes joining linear structural paper components.* Paper presented at the ICSA.

Octatube. (2010). Cardboard structures. Retrieved from https://www.slideshare.net/booosting/booosting-27mei2010-octatubecardboard2

Schütz, S. (2017). *Von der Faser zum Haus*: bauhaus ifex research series 1.

TU-Delft. (2008). *Cardboard in Architecture* (Vol. 7): IOS PRESS, Amsterdam, NL.

Voigt, P. (2007). *Die Pionierphase des Bauens mit glasfaserverstärkten Kunstoffen (GFK) 1942-1980.* (PhD). Bauhaus-Universität Weimar, Retrieved from https://www.yumpu.com/de/document/read/18589109/dokument-1pdf-19221-kb-bauhaus-universitat-weimar/206

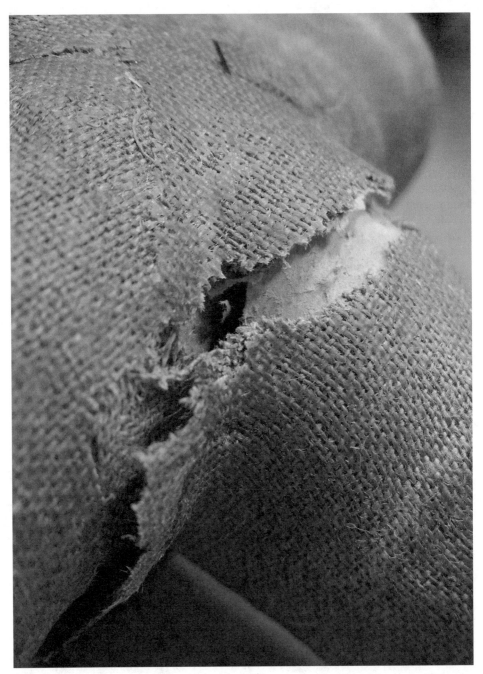

Figure 4.1 Failed specimen, case study 'Textile-reinforced epoxy resin laminated joint' (see paragraph 4.2.2.2).

4 Case studies

Keywords: Multi-axial connection, concept, design, detail, assembly, prototype, structural

In this chapter the experimental research on a variety of joining techniques for beam-structures composed of paper-tubes is presented. The aim is to generate answers for the third research question (and its sub-questions): 'Which joining techniques show the best potential for small scale tubular structures, to be used as a temporary living space, considering the aspects of design, assembly, stability, materialization and production?'. The environment for these experimental investigations, on multi-axial connections, is already described in paragraph 3.3.2 'Selection of specific context for the case studies'. Yet, in section 'guidelines', in the start of this chapter and also paragraph 4.2, the perspective of the investigations is further clarified.

At first the decisions for the selection of case studies are explained in paragraph 4.1 'Design concepts and selection of case studies'. The collection of design concepts, presented in table 4.1, that are viewed as a starting-point for this research process, is generated based on the pre-work developed as part of the 'state of the art' (chapter 2). The selection criteria are presented in table 4.2. Then the methodology used to proceed with the case studies is explained, in paragraph 4.2, 'guidelines'. The research process and findings, for each case study, are presented extensively in sections 4.2.1 'Plug-in joints' and 4.2.2 'Sleeve-joints'. For each case study three main areas are investigated. The first one includes the design and detailing alternatives that are possible within a design concept and the assembly method, also on a global scale (building method). The idea is to discuss the main features of the joining technique that influence its functionality and possible transformations that could make it suitable for implementation in various configurations. The second one is the manufacturing process, according to the methods used as part of the prototyping experiences and further possibilities with potential benefits for systematic production. The third one is the structural performance that is examined with two methods, structural testing and FEM simulations.

Global observations, further interpretation of the findings and comparisons between the joining techniques studied are reported in 4.3 'Evaluation of case studies'. The evaluation is mainly divided in two areas: the design development and the investigations on structural performance. This division serves the purpose of distinguishing the findings that are created with a 'trial and error' approach and are to be evaluated mainly in a qualitative manner, from these that include quantitative data as well, meaning the results provided by the structural tests and FEM simulations performed. The influences between the two levels of evaluation are considered important and thus efforts are made to analyse them, highlight the cross-links and suggest how further development could be achieved.

E. Kanli, *Experimental Investigations on Joining Techniques for Paper Structures*, Mechanik, Werkstoffe und Konstruktion im Bauwesen 60, https://doi.org/10.1007/978-3-658-34501-3_4

Guidelines

In this section the following aspects are clarified:

- Identification of main categories as 'principle joining methods' under which all design concepts occur.
- Comparability of case studies
- Reference systems used to address different features of the joining methods within the following case studies.

Specific information about the research tools and processes used within the case studies are presented in the start of paragraph 4.2.

Principle joining methods

Hereby the grounds on which the categorization of the design concepts is based are provided. Considering typologies of connections between beams, in this case the paper-tubes, a general distinction can be made between direct joints that mainly include examples of form-locking, occasionally secured with mechanical fixations and indirect joints such as pressure plates and intermediate nodes made of other materials like wood or steel. Based on the findings presented in chapter 2 'Joining techniques – State of the art', the main categories of joints selected for the elaboration of design concepts are:

- Connections based on form-locking: Simple and hybrid solutions are included where reinforcements apply, such as adhesives, ties or fasteners.
- Plug-in intermediate connectors: These can be divided either per material (steel, timber, composites (casted/ pressed), lightweight aggregate (casted), polymer - connector etc.)
- Sleeve intermediate connectors: This category includes thin-walled materials such as steel, aluminum, bio-based composites etc.

Comparability of case studies

All design concepts aim to provide solutions for the same design problem, as set in paragraph 3.3.2 'Selection of specific context for the case studies'. The aspects of detailing, assembly techniques, production method and structural analysis are considered as part of the elaboration of all case studies. Efforts were made to follow consistent methods while examining the different cases. In paragraph 4.3 an extensive evaluation is provided.

Reference system

When analyzing joining techniques and particularly multi-axial joints, it is essential to identify ways to refer to the different directions and groups of assembled parts. Figure 4.2 presents an example of a reference system that is inspired by the 'cartesian system of coordinates' suggested in 'VDI' (2004, p.45[1]), in order to describe the whole matrix of a joining system. There, the suggestion is to address the movements prevented by the joint, in all directions and between all elements using a code system, expressed in coordinates.

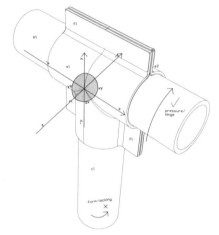

Figure 4.2 The elements of a joint and the relation between these. The joining method refers to case study 'Fiber-reinforced pulp-composite connector' (see paragraph 4.2.2.1).

In the following case studies simple methods are used for this purpose, as explained below:

- For the presentation of the assembly method and detailing options the characteristic design adjustments in different axes are addressed by name, per study.
- To report the experimental structural testing process a reference system is considered that follows the testing set-up and specifically the direction of the force in relation with the geometrical features of the joining system.
- Cartesian reference system is used to perform the local FEM analysis, in the environment of 'ANSYS Workbench' software. This is particularly useful to both provide input, also from the global FEM analysis (Dlubal RFEM software, see chapter 3.2) and also to discuss the results, with focus on the concentration of stresses.

---

[1] Methodical selection of solid connections. Systematic, design catalogues, assistances for work, Fachbereich Konstruktion. Ausschuss Verbindungstechnik.

4.1 Design concepts and selection of case studies

4.1.1 Design Concepts

A showcase of design concepts that could become the starting point for the elaboration of case studies in this research is shown in table 4.1. These concepts are inspired by the fin-dings presented in chapter 2 'Joining techniques - State of the art' and are placed in the three categories of form-locking, plug-in and sleeve joints, as these are described above, in paragraph 'principle joining methods' (guidelines).

The central aim of generating this showcase is to stimulate the discussion regarding the direction towards which this research is developing and set a milestone in this process, through a qualitative evaluation (tables 4.2 - 4.5). The qualitative evaluation is an opportunity to highlight the interaction between the concepts viewed in the showcase and the criteria that are important for this research. In other words, it is a way to formulate the design process and explain the considerations within the decision-making, in the path of discovering the research subject. At the same time, following the context set by the end of chapter 3 'global structural analysis', along with the outcomes from the FEM analysis, a design problem is set. Hereby, this is an opportunity to address the design-perspective in the research process, by sketching a variety of potential solutions. This approach is commonly regarded as 'drafting', in design and construction projects and is used as an instrument to communicate different options, until a suitable one is achieved. Only in re-search proving the functionality of a concept is commonly a longer and uncertain process. Still, this collection of concepts can be revisited as a source of inspiration.

Below the suggested concepts (table 4.1) are briefly discussed.

Form-locking based joints (concepts 1-3), mainly inspired by bamboo structures, involve simple form-fitting, cut-outs designed to prevent movement in certain directions and combination with simple mechanical fixations to increase stability.

Plug-in joints are extensively discussed in paragraph 2.2.4.1 'Hybrid techniques and in-termediate joints'. The designs include various materials, such as timber, steel and po-tentially composites. The joining mechanisms and assembly methods differ significantly. A common characteristic is the increased contact surface between the elements of the node and the paper-tubes, to achieve effective distribution of stresses and prevent shear-ing or rotation.

Sleeve joints form a protective skin outside the paper-tubes that could potentially be a composite material. Comparing to the previous two types of joints they are expected to cause less harm on the paper-tubes, as these could remain intact and also the implemen-tation of fasteners could be avoided.

Table 4.1 Design concepts for paper-tube beam structures.

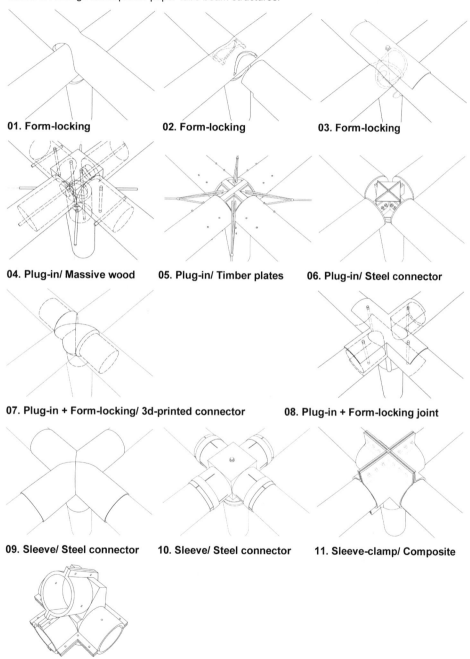

01. Form-locking

02. Form-locking

03. Form-locking

04. Plug-in/ Massive wood

05. Plug-in/ Timber plates

06. Plug-in/ Steel connector

07. Plug-in + Form-locking/ 3d-printed connector

08. Plug-in + Form-locking joint

09. Sleeve/ Steel connector

10. Sleeve/ Steel connector

11. Sleeve-clamp/ Composite

12. Sleeve-clamp/ Composite

4.1.2 Selection of case studies

For the selection of design concepts for further development the criteria expressed in table 4.2 are prioritized, based also on influences routed by the global aims and requirements (chapter 1, research methodology), as following:

Joints are often characterized as the weak links in a skeletal structure, with a crucial role in the overall stability. Within the boundary conditions as these are set in chapter 3, it is important to search for joints that would perform as a strong link between the beams.

Considering the general requirements for developing structures that can be realized rapidly, efficiency in production is important, as well as avoiding high-effort assembly processes. Next to this, the contribution of the joints in the reversibility and recyclability of the skeletal structure is important, as part of the global aim. Furthermore, it is useful to aim for joints that present durability that is similar to this of the main structural elements and not necessarily much higher.

Further than the criteria addressed in table 4.2, the possibility to prefabricate structural modules is an important aspect. The majority of design concepts addressed hereby are compatible with this. Still, parts of the joints, such as adhesive bonds are susceptible to damage during transfer and thus, further measures need to be developed.

Based on the findings presented in chapter 2, plug-in connectors, such as concepts 4 and 5 of table 4.1 are already presented multiple times in reference projects. The use of materials with wood origins is particularly attractive as a solution to complete a paper-based structure. Moreover, the easiness in production, as a result of simple forming methods that don't require any high-technology means is another reason that makes such concepts attractive. However, there is very limited information regarding technical aspects and particularly the structural performance of these types of joints. On these grounds, further investigation of these concepts is considered particularly interesting.

Overall, plug-in joints help to increase the contact surface and thus optimize the distribution of stresses between the adjacent beams. Simplifying the design and production further (comparing to concepts 4 and 5) and allowing for minor errors in the assembly are characteristics met by concept 8.

In a similar way, sleeve joints also help to optimize the transferability of forces, on a form-locking based beam to beam connection. However, most known examples regard metal connectors. Investigating design concepts 9 and 11, with the aim to develop thin-walled joints realized with bio-based materials is an interesting alternative.

To conclude, the design concepts 4, 5, 8, 9 and 11, from table 4.1 are selected for further development. As part of the design cases discussed above, good grounds are created in order to examine certain types of fasteners and adhesives as parts of hybrid joints.

Table 4.2 Qualitative evaluation method and scale (see accompanying legend below).

Design concept	Transferability of Forces				Design							Durability		Recyclability
					Production		Cost		Assembly / Dis-					
	P^2	T	S	M	A	H	M	P	A	D		0	1	
Nr. X														

Legend (in principle the colour-scale aims to indicate positive characteristics with lighter tones)

Transferability of Forces

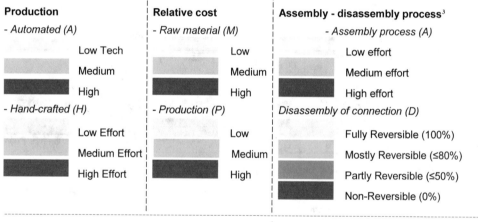

High

Medium

Moderate

Poor

Production

- *Automated (A)*

Low Tech

Medium

High

- *Hand-crafted (H)*

Low Effort

Medium Effort

High Effort

Relative cost

- *Raw material (M)*

Low

Medium

High

- *Production (P)*

Low

Medium

High

Assembly - disassembly process[3]

- *Assembly process (A)*

Low effort

Medium effort

High effort

Disassembly of connection (D)

Fully Reversible (100%)

Mostly Reversible (≤80%)

Partly Reversible (≤50%)

Non-Reversible (0%)

Durability

Durability of the connection (0)

High

Medium

Low

Durability of the connection in comparison with the beam component (paper-tube) (1)

Significantly higher

Higher

Similar

Lower

Recyclability

Fully Recyclable

Mostly Recyclable

Partly Recyclable

Non-recyclable

---

[2] 'P' stands for pressure, 'T' for 'tension', 'S' for 'Shear' and 'M' for 'Moment'
[3] 'effort' is translated to time, tools and skills

As explained in the start of this paragraph, per 'design concept' certain criteria are considered, following the global aims of this research, as part of a qualitative evaluation process that leads to the selection of concepts for further development. These are expressed in table 4.2 and the qualitative evaluation is presented in tables 4.3 - 4.5.

Table 4.3 Qualitative evaluation of design concepts based on simple form-locking.

Design concept	Transferability of Forces				Production		Cost		Assembly / Dis-		Durability		Recyclability
	P	T	S	M	A	H	M	P	A	D	0	1	
1		*								**			**
2		*								**			**
3										**			**

*Complementary fixation mechanism is required (for example ties).

** The reversibility and recyclability factors depend on the implementation of adhesives.

Table 4.4 Qualitative evaluation of design concepts of 'plug-in' connectors.

Design concept	Transferability of Forces				Production		Cost		Assembly / Dis-		Durability		Recyclability
	P	T	S	M	A	H	M	P	A	D	0	1	
4		*											
5		*											
6		**		**	-								
7***						-							
8					-								

* Bolted or screwed mechanical fixation is required.

** For the fixation of the connector in the paper-tubes steel rods can be implemented to pre-stress the tubes and ensure high performances (see paragraph 2.1.2.2, fig. 2.38).

*** Similar designs can be produced with casting instead of 3d printing. In that case a wide variety of materials can be considered (such as reinforced concrete, bio-based composites etc.).

Table 4.5 Qualitative evaluation of design concepts of 'sleeve' joints.

Design concept	Transferability of Forces				Production		Cost		Assembly / Dis-		Durability		Recyclability
	P	T	S	M	A	H	M	P	A	D	0	1	
9					-								
10					-								
11						-				**			**
12						-						*	

* The durability depends highly on the materialization. For instance, if such a connector is attempted with pulp-based mixture, for casting, the durability will be low. On the contrary, fiber reinforced composites could lead to more durable solutions.

** The reversibility and recyclability factors depend on the implementation of adhesives.

4.2 Case studies

Guidelines for the elaboration of case studies

Hereby the methodology followed for the development of the case studies is presented, emphasizing on the studies about structural performance. The aim is to present the approach and purpose of the studies and to describe the bigger picture, as well as the tools and evaluation methods. Every case study starts with the description of the joining method, characteristic details, suggestions and experiences about the manufacturing process, in a straightforward manner. The next part that is related to the structural performance presents higher complexity. Therefore, certain clarifications about the common conditions that apply for all studies are necessary. These mainly regard the processes of experimental structural testing and local FEM analysis, focusing on the data extracted and the further processing of these to obtain further conclusions. Differences in the depth of elaboration between some of the case studies appear and are related to either limitations in the accessible manufacturing processes or shortage of input (data) or time.

Development of the design

For each of the case studies elaborated, a principle design and assembly method are used as a starting point, defined as 'characteristic joining method'. In the respective paragraphs, further possibilities are discussed, in relation with design adjustments that could serve different purposes. The detailing alternatives considered may differ, depending on the original joining method. To give an impression for the possibilities considered in this research, a few examples are mentioned. A possible reason could be changing the structural behavior, either by strengthening the joint itself or by adjusting the contact-relation with the paper-tubes. Another approach could regard alternatives about the implementation of different fixations between the joining elements (for example pressure ties instead of fasteners). In a different way, if there is a special focus on prefabricating certain units, the design of the joint could support a specific assembly-sequence. Furthermore, changes in the design could be required to support certain manufacturing methods.

Prototyping

This section is of great importance for this research, due to the experimental character and the approach of building prototypes to test the structural performance. This way, the experimentation and manufacturing methods followed for the creation of prototypes are the main subject. The advantages and the difficulties met in the different case studies are addressed and potential alternative manufacturing processes are discussed.

Experimental structural testing

Various testing methods were considered. The performance of the joint for shear, bending, tension, pressure and torsion (this last one especially due to winding of tubes) are all important. Yet, only a selection of experiments could be incorporated in this research, based on the available equipment and timeframe.

Some of the most interesting testing methods for beam-joints reviewed prior to the development of experiments are the following:

1. Studies on common Japanese longitudinal joints for timber structures, where the testing set-up is similar to the present experiments (4-point bending test) (Kohara, 2004).

2. Studies on the 'moment-rotation behavior of beam to column joints' (Mottram, 2012), for steel bolted beam joints between I-sections.

3. High-tech concepts that could also apply for testing of multi-axial nodes. The challenge of testing multi-axial joints lies within the difficulty to create a combination of loads with a classic device for structural testing. An impressive example of equipment that has been created to test components for ships regarding their fatigue behavior, with vibrations, is 'Hexapod'[1], a machine developed by TU Delft.

As a main priority for testing, the method of four-point bending tests was selected, mainly for three reasons.

*   In principle, a paper tube is weak for bending (see example of testing results below). Next to this, for skeletal or space structures composed of paper-tubes, the effect of bending appears to be the biggest problem that leads to buckling of the structure, as it is also indicated by the global structural analysis (chapter 3).

*   As it has been observed in the global structural analysis, the bending stiffness of the joint is very important for the stability of the structure.

*   For the majority of case studies developed, the effect of bending is the most difficult to prevent. The reasons for this are partly related to the round, hollowed shape of the beam, in combination with the material behavior that make it more challenging to create a suitable base for the connection.

Therefore, the performance of the 'case studies' for bending is important to quantify, monitor the issues, the failing mechanisms and also identify if some joining methods are more effective.

---

[1] A summary for this technology can be seen here: https://www.tudelft.nl/en/3me/about/departments/maritime-and-transport-technology/research/ship-and-offshore-structures/facilities/hexapod/ - (accessed 12.2020).

Further testing methods tried as part of the experimental research process are the following:

- Experimental vertical bending test, in the same spirit as the second testing method, as reviewed above.

- Experimental torsion test (following the previous testing method).

- Axial compression test, to create a shearing effect between the connected elements.

- Shear test to examine basic assemblies of bolted, stapled and glued joints.

Guidelines for the bending tests

- Testing rig

The testing rig is presented in figure 4.4 that presents a paper-tube being subjected to the test, as a first example. To describe the main conditions for this test, the steel frame provides two supports at the bottom and two points for application of load at the top. Most commonly, cylinders are used to support the specimen directly. In this case, the angular supports that are fixed on top of the cylinders, were especially designed to improve the contact with the specimen (Kiziltoprak, N.), following earlier experiments with testing of paper-tubes. Curved profiles instead of angular ones would be the optimal choice. At the locations where the force is applied, the Force and respective displacement are monitored.

The speed for applying the force on the specimen was set as 'position controlled', with a rate of 5mm/min. Regarding the climate conditions in the room where the tests were performed, the temperature was about 22°C and humidity 33% RH approximately.

- Measurement of vertical displacement in the middle with a sensor

For the final series of testing, a sensor is placed in the middle of the rig (fig. 4.3), to monitor possible vertical displacement other than that measured on the sides, at the top supports. To explain this, as the assembly of the specimens consists of three elements, the middle area, where the connection between the two side-beams is realized, might develop slightly different vertical movement. By quantifying such differences, concrete conclusions can be drawn for the behaviour of the joint, as a reaction to the forces that push the specimen downwards. When the monitored values are positive, this means that the middle point of the specimen moves further downwards. In the opposite way, when the value is negative, then the middle moves upwards. In table 4.8, the values monitored for the different series of experiments, for $F_{max,Linear}$ and $F_{max,Ultimate,}$ are presented, with the aim to provide a global impression about the displacement measured by the sensor, for each case study. Per case study, the range of values monitored for all repetitions is used to describe the minimum-maximum margin for the above values. It can be easily understood that monitoring the displacement in the middle is not as important for all of the case studies. However, in some cases it provides useful insights, such as the testing series 'timber-block node, configuration 2' (see 4.2.1.2, structural testing).

Table 4.6 Spectrum of displacement (min. - max.) measured by sensor during the 4-point bending tests, placed in the middle of the distance between the supports, as shown in figure 4.3.

Testing Series (Code- name)	Number of repetitions	dl_B min - max (F_{max} Ultimate) (mm)	dl_B min - max (F_{max} Linear) (mm)
Paper-tube	3	1.09 – 1.29	0.21 – 0.77
Timber plate puzzle node, configuration 1 (see paragraph 4.2.1.1)	6	-0.10 – 0.94	-0.66 – 0.85
Timber plate puzzle node, configuration 2 (see paragraph 4.2.1.1)	3	-0.01 – 0.64	-0.45 – 0.14
Timber plate puzzle node, configuration 3 (see paragraph 4.2.1.1)	3	-9.56 – -5.63	-6.5 – -2.98
Timber block node, configuration 1 (see paragraph 4.2.1.2)	3	1.82 – 3.38	0.65 – 1.39
Timber block node, configuration 2 (see paragraph 4.2.1.2)	3	-17.63 – -16.80	-5.68 – 0
Textile-reinforced epoxy resin laminated joint, configuration 1 (see paragraph 4.2.2.2)	3	0.48 – 0.73	0.19 – 0.29
Textile-reinforced epoxy resin laminated joint, configuration 2 (see paragraph 4.2.2.2)	3	-0.38 – 1.56	0.27 – 1
Tolerance adaptive timber plug (see paragraph 4.2.1.3)	1	0	0

Processing the results from structural testing

- Selection of representative values

The aim is to provide insights about the performance of the specimens and to create useful input for the evaluation. For each series of experiments the average graph-line has been created. This graph is used to identify some values that are selected for representing the performance of the specimen. These are mainly the maximum force that the specimen can carry before the plastic deformation begins ($F_{max,Linear}$) and the maximum force before the specimen loses strength and fails ($F_{max,Ultimate}$), accompanied by the vertical displacement at these points.

For the definition of F_{max} Linear, a linear curve is defined, like in graph 4.2. The $F_{max\ Linear}$ is selected on the higher limit of the overlapping area, with the actual curve of the specimen. A margin of 2° for the angle between the two curves has been decided, as a limit for the deviation between them.

This approach has been preferred due to the limited amount of experiments. Otherwise, according to existing standards (The European Union Per Regulation 305/2011, 2005, p. 108) a statistical distribution 5% is suggested.

- Evaluation methods and criteria

The values selected above are used as input for further calculations. To be specific, the properties of the specimens that are calculated are the following:

- o Bending strength
- o Bending stiffness
- o Bending stress (Point B - top support)
- o Rotational stiffness y (Point B - top support)

The properties regarding the bending behaviour are related exclusively to the geometry of the tested specimens and cannot be generalized. Hence, they characterize only the specific specimens. Further details on the calculations are reported in paragraph 4.3.

Additionally, comparisons between the structural behavior of the joining methods examined and the paper-tube are made on the following grounds:

- o Bending strength of specimen vs. strength of tube
- o Bending stiffness of specimen vs. stiffness of tube
- o Comparison of failing mechanism

Structural behavior of selected paper-tube

Based on existing knowledge, in theory, paper works effectively when in tension, a principle that is expected to apply also for the element of the paper-tube. However, proving this would be rather complicated, especially due to difficulties in building an optimal testing set-up. Therefore, the limits of the material have been defined based on the results from axial compression tests (Kiziltoprak 2018). These properties are stated in chapter 3, table 3.2, as they were used as input for the global FEM analysis of typical structural cases.

Below the testing results of paper-tube specimens are presented (three repetitions), as an example of implementation for the process discussed above.

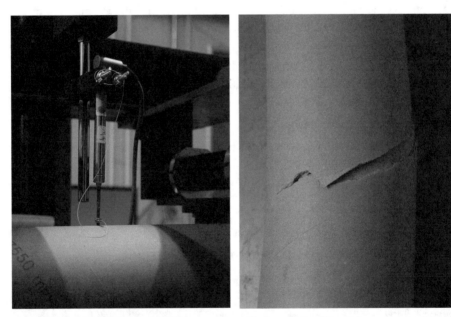

Figure 4.3 4-point bending test - (left) sensor placed in the middle of the distance between the supports, (right) failed specimen of paper-tube

As shown in fig. 4.4, when the tube fails, the crack is initiated on the bottom side, right under the support that applies pressure on the top of the tube, causing tension along the bottom line. The curve that the crack develops is parallel to the winding line, which is the basis for the production of the component.

Figure 4.4 4-point bending test - (top) testing rig, (bottom) failed specimen

Table 4.7 Testing results from 4-point bending test of paper-tubes.

Configuration 1 Specimen Nr.	F_{max} Ultimate (N)	Displacement Point A (mm)	Displacement Point B (mm)	Time (sec)
Specimen 1	7484	23.04	1.29	277
Specimen 2	7322	28.44	1.09	342
Specimen 3	7484.5	24.16	1.21	290

Table 4.8 Representative values selected accordining to the provided guidelines (paper-tube).

Paper-tube	Mean F_{max} Ultimate (N)	Displacement Point A (mm)	Mean F_{max} Linear-part (N)	Displacement Point A (mm)
	7369	21.5	6100	11.7

Graphs 4.1 Results from 4-point bending test. - Specimen: paper-tube (inner diameter 100mm, wall thickness 10mm)

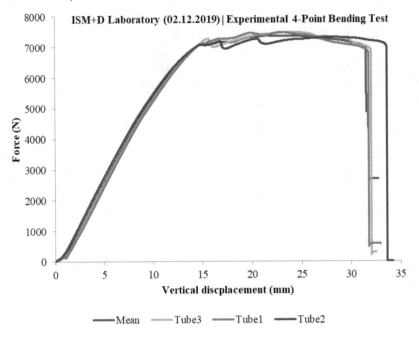

Graph 4.2 Mean curve, extracted from the 3 repetitions (see graph 4.1) and selection of F_{max} Linear.

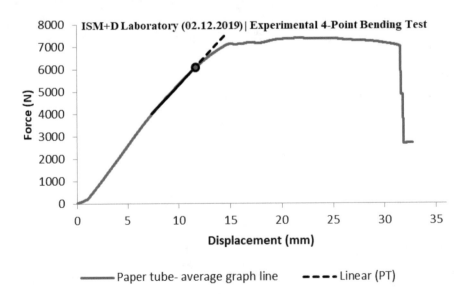

Local FEM Analysis

The software selected for this purpose is 'ANSYS Workbench (Static Structural)'.

The main goal is to analyze the distribution of stresses within the joining system. Then, the resulting stresses are also compared with material properties, to report potential failure on critical areas.

- Simplifications within the modelling part

Important simplifications were made for the definition of material properties, by assuming linear elastic behavior for the following materials and selecting as input the properties of the weakest direction: paper tube, plywood, timber beams. Further details are mentioned in the case studies.

- Material properties that apply for all case studies
- o Paper tube: Young's modulus 728 MPa and maximum stress 9MPa (Kiziltoprak, 2019). For the maximum stress, the resulting stresses should be significantly lower than the value provided here, considering the idealization of the component, as modelled in ANSYS. The Poisson's Ratio applied is 0.2 following the testing results from the project 'Japan Pavilion' (McQuaid, 2003)
- o Structural Steel 200000 MPa and Poisson's ratio 0.3, max. stress 235MPa (Material Standard EN 1025-2)
- o Sliding friction value, between the paper-tubes and wooden elements, as well as between wooden elements is set to $\mu=0.4$ (Nihat Kiziltoprak, 2019)

Stress-analysis of multi-axial nodes

- Input

The structural concept as well as the Forces and Moments to be applied on the geometries are already described in paragraph 3.3.2, table 3.11.

Local co-ordinate systems are applied in the ANSYS models, following that structural model. Further details are provided within the different case studies.

Regarding the contact properties between all parts of the geometries, several possibilities were examined. Through this process the ANSYS models were optimized in order to create the desired effect regarding the distribution of stresses. Specifics are reported per study.

Overall, the desired load combination is not possible to perform with structural testing within the current studies and therefore the ANSYS simulation is the alternative method used instead.

- Output

The main simulation tool selected, to examine the distribution of stresses in the system, is 'principle stress analysis'. Deformation analysis is performed but cannot be directly compared with the results from the global FEM analysis. This is, primarily, due to the different boundary conditions, such as the method for applying loads. To a further extend, also the difference between the simply hinged joints that are applied in the global analysis, comparing to the designed joining systems examined in this section, lead to different deformations.

Stress-analysis of linear configurations

The models analysed represent the specimens of the previous paragraph (experimental bending structural tests, see example in fig. 4.5). Again, stresses and deformation analyses are performed.

The resulting stresses can be compared either with these calculated for the same specimen or with the maximum stress that the materials can withstand. In the first case, the aim is to get a glimpse on correlation between the results obtained from structural testing and the ANSYS simulations. In the second case, more global conclusions are obtained, by understanding the critical areas and failure modes encountered by ANSYS. This way, useful feedback is provided both for errors within the testing process, such as imperfections that might worsen the results, or errors in the set-up of the FEM analysis.

As within the software at least the geometrical conditions are idealized (0 imperfections) and the mechanical behavior of materials is simplified, it is interesting to compare the results from both sides. The forces applied are in total equal to the F_{max} Linear, as calculated for the different cases.

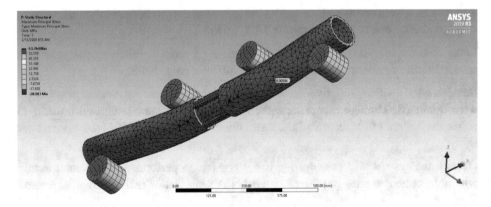

Figure 4.5 Stress analysis of joining system in ANSYS software (case study 4.2.1.1).

Literature

Kiziltoprak, N. (2019). *Experimental method to determine the bending load capacity of paper tubes.* Retrieved from https://www.ismd.tu-darmstadt.de/forschung_1/material_ismd/bamp____bauen_mit_papier/bamp___bauen_mit_papier.en.jsp

Kohara, K. (2004). A study on experimental testing of joints on timber structures. *13th World Conference on Earthquake Engineering, Vancouver, B.C., Canada*(Paper No. 2441).

McQuaid, M. (2003). *Shigeru Ban.* London N1 9PA: Phaidon Press Limited, Regent's Wharf All Saint's Street.

Mottram, J. T. (2012). *Moment-rotation behaviour of beam-to-column joints for simple frames of pultruded shapes.* Retrieved from

Nihat Kiziltoprak, E. K. (2019). *Load capacity testing method for non-conventional nodes joining linear structural paper components.* Paper presented at the ICSA.

The European Union Per Regulation 305/2011, D. E., Directive 2004/18/EC]. (2005). EN 1990 (2002). Eurocode - Basis of structural design. In.

4.2.1 Plug-in joints

In this section, three different cases of joints are studied. The first one mainly regards a design identified by timber plates that form a 3d puzzle, whereas in the second case timber blocks are used instead. In the third study, an attempt to develop a solution that could be more effective for absorbing tolerances is made, by combining the principle of form-fitting between the tubes with simple joining elements.

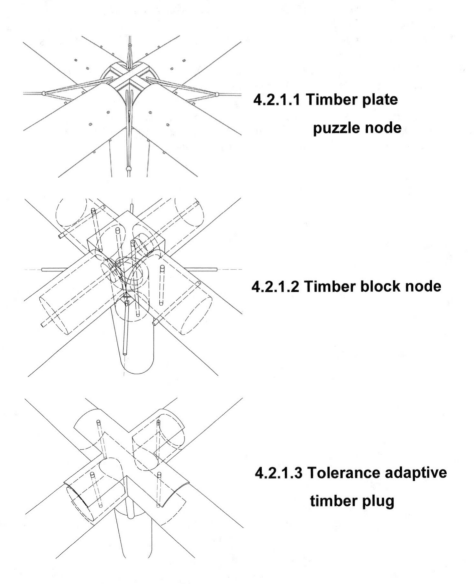

4.2.1.1 Timber plate
   puzzle node

4.2.1.2 Timber block node

4.2.1.3 Tolerance adaptive
   timber plug

4.2.1.1 Timber plate puzzle node

Figure 4.6 Prototype (07.2019) – integration of slits and stiffening corners.

This joining method has multiple references in existing studies. Examples of application are referenced in paragraph 2.1 'Collection of reference projects and studies', table 2.1, project numbers: 2, 4, 5. Representative construction details for joining methods of this kind, based on the respective references, are provided in paragraph 2.2.4.1 'Light timber nodes'. Specifically, in fig. 2.29 the basic types of interlocking between the plates are demonstrated and in fig. 2.31 an example of a multi-axial roof joint is shown. Based on the findings, this technique is highly preferred due to the easiness in production that has benefits for the erection of emergency shelters. At the same time, there is very poor evidence for the structural performance. In this respect, the wide variety of wood products that are used to build these nodes is also an issue. Within this research, the first experimental case study 'House 1' (for further information and details see paragraph 3.1.1, fig. 3.4 and appx. 6.1), also makes use of this joining principle.

The goal of this case study is to examine further the assembly method of this joint, develop an optimized version and create proof for its functionality, especially about the aspect of structural performance. For this purpose, some important factors are related to the quality of the timber plates, the fixation between the joint and the paper-tubes and the manufacturing process. The greatest challenges are linked to the different structural behavior of the joints in the three main axes and the integration of mechanical fixations, between the joints and the tubes.

Further studies on this assembly method were conducted as a preparation for the 'Branch-out' pavilion that was erected at the exhibition 'Glass technology live' in Messe Dusseldorf, date: 10.2018 (see appendix 6.3).

Table 4.9 Characteristic joining method – 'Timber plate puzzle node'.

Form-locking			Plate to plate puzzle Timber-cross to tube
Force-closure	**Friction**		Contact area between plates Timber joint with tube
	Pressure		Screwed connection between timber joint and tubes
	Tension		Timber connector Screwed connection
	Inertial Forces		Tensile cables
Material-closure			Corner reinforcements glued on timber node

Detailing alternatives

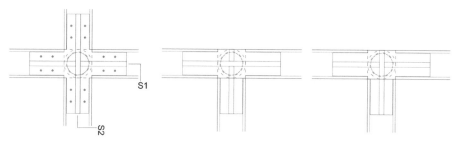

Figure 4.7 (left) Horizontal section, (center) vertical section S1, (right) vertical section S2

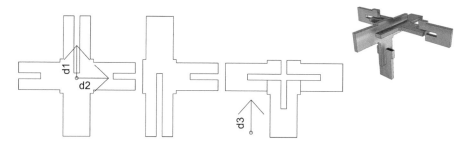

Figure 4.8 Directions d1 and d2 are on a horizontal plane, whereas d3 on a vertical.

Design of plates

For a steadier joint, it is advisable to integrate as many axes as possible on the same plane. Next to this, 3d-puzzle designs such as joining technique 'lap form-lock' nr. 6, table 2.5 (paragraph 2.2.3) are possible, especially when the thickness of the plate is high, to develop 3d form-locking mechanisms (then milling instead of laser-cutting is the process required to form the plates). To strengthen the joint, the design of the plates around the perimeter can be adjusted, in order to reinforce the corners by expanding the surface. Cuttings on the tubes, otherwise named as slits (see fig. 4.13), might be required for integration of wide parts, as reinforcement.

Fitting with paper tubes

When the edges along the perimeter of the plates are not treated and are therefore sharp, the inner surface of the tube is locally torn. Curving the edges of the plates helps to increase the contact surface with the tubes (in case of CNC Milling, the diameter of drill creates a curved edge on the one side of the plate). Next to this, to ease the assembly process, the endings of the joint can be designed to allow for slightly higher tolerance (about 2 - 5cm distance from the edge). The integration of slits could in theory improve the form-fitting with the tubes, with positive effects against rotational movement caused by torsion. This possibility shall be tested to quantify the effect. Within this research, bending tests were carried out, from which the conclusion for this aspect is that the performance of this type of joint is not significantly affected by the presence of slits, under the condition that wooden reinforcements are applied at the inner corners of the cross-joint (fig. 4.13 and appx. 6.2). The same wooden reinforcement may also be used as a spacer between the tubes, to increase performance against shear forces.

Mechanical fixation

The wooden reinforcements are used to secure the joint in the tubes with wood screws (diameter 4.5, depth 35 mm). A distance of 35 – 40 mm is kept between the screws. Structural testing (see page 212) proved that the glue bond between the reinforcements and the timber cross is stronger than the mechanical fixation, a conclusion that is a positive indicator for this assembly method. Otherwise, without the reinforcement, the cross could be bolted with the tube, in a direction perpendicular to the surface of the plates, but especially due to the round shape of the profile and the empty space this would not be convenient. At last, screwing the cross with the tubes along their shared surface might be the easiest, yet the most unsafe option, due to the unfavorable fiber orientation. This option was tested prior to the construction of the 'Branch-out' pavilion (appx. 6.3).

Tensile elements

Flexible cables or thin rigid rods can be attached, to stiffen the structure. In this case, the better option would be to create space for their integration on the main plates of the node. The fixation point would be different for each of these two cases. On the positive side, such an addition does not occupy much space nor does it imply significant complexity overall. On the other hand, local concentration of stresses due to excessive tensile forces can cause the timber plate to rupture. Thus, minor adjustment of the design or local reinforcement of the material might be useful.

Design flexibility

Regarding the material selection, relevant information is stated in the section about production and prototyping. About the translation of the characteristic joining method to designs with different number of axes, depending on the construction outlines, the current technique provides relatively good possibilities. In theory a maximum of six tubes can be connected with this joint. More axes would possibly lead to excessive concentration of stresses at the center of the joint that exceed the capacity of the material. Thus, structural symmetry, within the global structure and also the joint itself, is important for the functionality of the joint.

Assembly techniques

A common approach is to form the individual joints and then assemble the structure with sequential order (from the bottom - fundament - to the top and from side to side). On-site assembly of the structure is a realistic option in this case, especially when there is a limited number of structural elements with different geometrical characteristics. In a different way, prefabrication of units is also possible. The structural system is basically always designed to serve the assembly process. For the present structural system, the assembly method implemented for 'House 1' (3.1.1, fig. 3.4) presents interest. The structure is divided in planes that can be pre-assembled and then erected and fixed together (fig 4.9).

In principle, the method of inserting the joints in the tubes depends on the tolerances left between the tubes and the joint. Normally just a rubber hammer should be sufficient. The smaller the tolerance the more effort and time is required. Another option would be to use machinery to push the joints in the tubes by means of pressure. However, this solution is more likely to be needed for massive wood plug-in joints (4.2.4.1, 'Massive wooden nodes').

In principle, in case parts of the structure are prefabricated, for the process of transportation, it should be ensured that any forces, despite the own weight, that might be acting on the units are eliminated, to avoid possible damage of individual elements or at the areas of mechanical fixations etc. Relocation of the structure is not highly recommended, as the structure will possibly need to be disassembled for this and the screwed joints won't work as effectively when fixed for a second or third time. Still, there are measures that could support this idea, such as designing a few reusable joints between prefabricated units, using slightly bigger screws for a second assembly etc. Additionally, there are further factors with impact on this possibility, such as the condition of the structure and other boundary conditions.

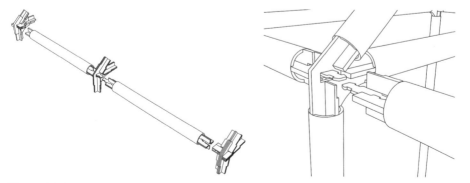

Figure 4.9 'House 1', (left) assembly method between modular sections, (right) form-lock key-point.

Production and prototyping

Figure 4.10 Handcrafted joint (timber plate puzzle node).

The main steps for the production are the following: forming of single plates and wooden reinforcements, assembly of the plates, lamination of reinforcements, creation of slits on tubes if required. Some basic decisions include the selection of materials and cutting processes. The main possibilities are described below.

Forming the plates

The two options are either to hand-craft the plates or to use CNC-milling technology.

Some tips for handcrafting include the combination of basic wood-working tools for a fast and easy process. For instance, a sawing table can be used when possible and free-hand sawing to complete the cuttings at the inner corners. Before cutting close to the inner corners, round holes shall be drilled on the edge of these boundaries (either with a table-drill or a simple drill), big enough to fit the sawing blade, in order to ensure a clear corner-cut. At the end, the edges of the plate can be curved with a hand-router.

Figure 4.11 (Left) sawing table, (middle) free-hand sawing tool, (right) router

For the CNC Milling process a few basic rules apply for the digital drawings, such as minimum distance and incorporation of the diameter of the drill in the drawing where needed (inner corners). Moreover, as the elements to be cut are integrated in the perimeter of the board, correct placement according to the fiber orientation is possible and occasionally important. For this purpose, most often, the code-name of the product shows the fiber direction (where the dimensions of the timber-plate are stated, the first number indicates the superior direction). In the case of plywood, based on the number of layers, this point can be more or less significant. An important detail about the form-locking concept, is to leave tolerance of 0.25-0.5mm, as the wood might shrink shortly after cutting. With 0 tolerances the assembly process could be particularly difficult and disassembly would not be possible. To correct such errors, sanding the surface could help, decreasing this way the thickness of the plate. Next to this, in case corner-reinforcements are implemented, then the reversibility of the assembly is only possible by damaging the joint.

A downside of CNC Milling in relation with form-fitting is that the inner corners will present a smooth curve, as shown in fig. 4.12, that will prevent contact between the two narrow edges at the center of the cross. To resolve this issue further treatment of the corners (sanding) is required to shape 90° corners.

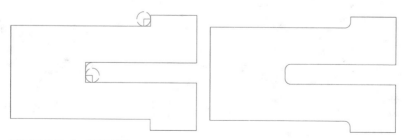

Figure 4.12 CAD file for CNC-Milling

Figure 4.13 Production of specimen for preliminary structural test (appx. 6.2).

Timber-plates

An important parameter for the strength, easiness in manufacturing and production cost is the type of timber plate selected. The type of wood plays a definitive role in the performance of the joint, as well as the thickness. A wide variety of products can potentially be used, with A-grade plywood being the product with the highest quality and also cost. The final prototypes that were subjected to structural testing were made of Multiplex. More specifically, the exact type of wood is Birch Plywood, BB (III), Standard grade that meets EN 635 requirements. Furthermore, the veneers are 1,4 mm thick and are cross-bonded. Experiments performed at an early stage of the research regard joints made of MDF board. Then, preliminary experiments indicated that MDF is not sufficiently strong.

Wooden reinforcements

For the corner-reinforcements, profiles that are slightly thinner than the inner corner of the cross, in this case 40*40mm, are needed. The specific type of wood selected is douglasie natural timber. Even though it's not easy to find profiles with the desired arched-edge, it is possible to approximate the shape with simple cuttings.

Lamination

Wood glue (Ponal) is used for the lamination of the reinforcements on the plates. Multiple clamps are placed diagonally for the curing process.

Creation of slits

There can be many ways to make these cuts. In this particular case, with the equipment available this is the process followed:

- Use of a sawing table for the most part
- Use of a custom-made support for the tube (fig. 4.14 - left)
- Finishing the inner corners with free-hand-sawing and sanding
- Application of glue along the open edges prevents splitting

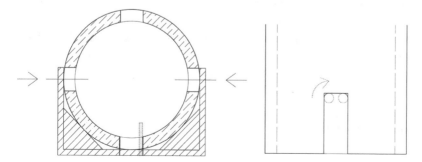

Figure 4.14 (left) Base support for tube, (right) cutting inner corner of tube for slit.

Structural Testing

Figure 4.15 Basic core of the specimens (final bending tests).

The performance of the joint in the three different directions, as these are addressed in the section about 'detailing alternatives', is examined (see figures 4.7 and 4.8). The main goal is to identify the limits of those when subjected to bending and observe the failure modes. In principle, critical areas occur close to the screws and the end of the joint, as the system there becomes rapidly weaker. A similar testing process is presented in the book Shigeru Ban (McQuaid, 2003, p. 79).

The concept of integrating slits as form-locking between the joint and the tubes is dismissed, as the prototypes showed that these don't create a significant effect and also weaken the tubes. Still, results from preliminary structural tests are presented in the appendix, paragraph 6.2 'Preliminary Experimental Bending Tests'.

Figure 4.16 presents the main outlines of the testing set-ups. As indicated there, the names 'configuration 1', 'configuration 2' and 'configuration 3' apply. These names are used to address the different tests, present the testing results (graphs and selected values) and to draw comparisons between the different configurations. Before the testing process, it is suspected that 'configuration 1' is significantly stronger comparing to '2' and '3' is expected to present significantly lower strength.

Furthermore, a series of compression tests is conducted to examine the effect of shear between the joint and the tube, with the aim to observe the failure mechanism and also ensure that the screwed joints fail before delamination between the wooden reinforcements and the cross occurs.

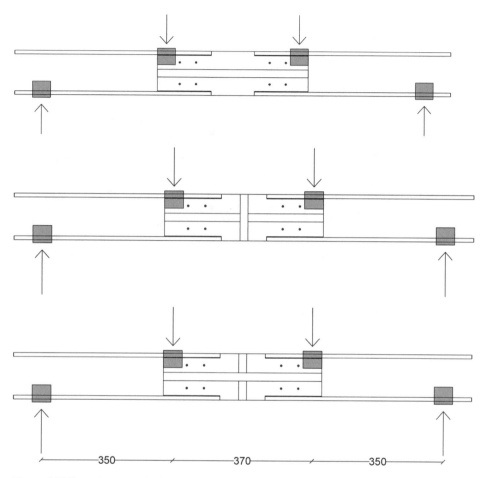

Figure 4.16 From the top to the bottom: testing configurations 1, 2 and 3.

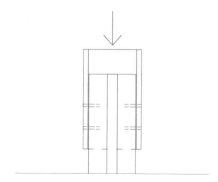

Figure 4.17 The set-up for testing the performance of the mechanical fixations against shearing.

Testing configuration 1

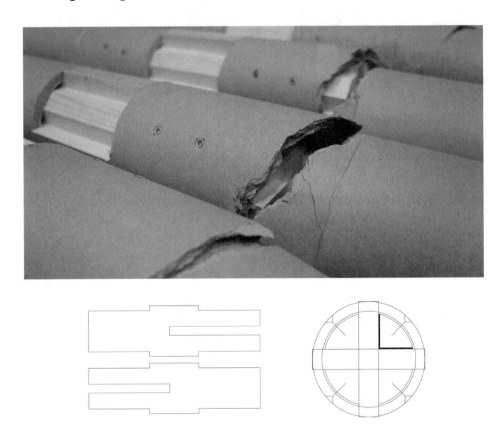

Figure 4.18 (Top) Failed specimens, (bottom) design of plates for 'configuration 1'.

In total six specimens were tested. All specimens failed in the same way, either on the side of support 1 or 2. The wooden joints remained intact and the displacement that is monitored by the sensor indicates slight changes in the position of the joint in a range of ± 2 mm. These basically represent the tolerances between the joint and the tube that were absorbed as the force rose and the structural system started bending. Since the joint is stronger than the paper-tube, the system fails at the area where it suddenly becomes weaker. Thus, all specimens failed close to the outer boundary of the joint and following the winding line (fig. 4.18, top).

The contact between the tube and the edges of the cross is limited. In principle, this fact could create various problems such as damage of the contact area and unfavourable conditions for the screws, especially for shearing.

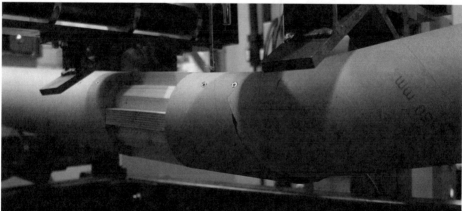

Figure 4.19 'Configuration 1' (top) testing rig, (bottom) failed specimen.

Table 4.10[1] 'Timber plate puzzle node' ,Configuration 1', testing results and selected mean values.

Configuration 1, mean values	Mean F_{max} Ultimate (N)	Displacement Point A (mm)	Mean F_{max} Linear-part (N)	Displacement Point A (mm)
	8253	36.7	6500	20.3

---

[1] The values presented in this table are selected from the average graph line (red line in graph 1)

Graph 4.3 'Configuration 1', testing results, force versus displacement curves. To explain the legend, the initials 'CW' were used as a code name for this case study, for the testing process, and 'c1' indicates the number of 'configuration 1'.

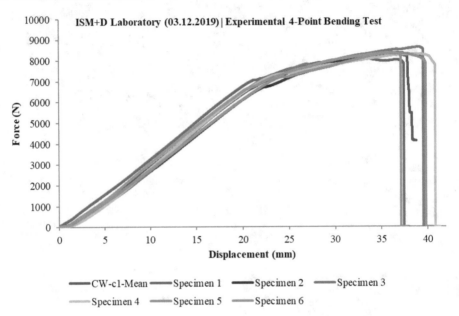

As it can be observed in graph 4.3, the force versus displacement curves follow the same pattern and also lie close to each other. Still, the margin between the maximum values is quite broad. More specifically, the Mean maximum ultimate Force is 8253 N with a margin of 527 N between the minimum and maximum value presented, whereas the Mean maximum Force, within the linear part of each curve, is estimated at 6545 N with a margin of 674 N. It is suspected that the reasons for the broad margins are mostly related to the matrix of the paper tube, so the winded layers as, in fact, each of the pieces of tube cut in order to form the specimens is unique.

Within the first 2 mm of vertical deformation approximately most curves indicate a slight deep. In principle, about 2,5 mm of deformation are expected as a result of the tolerance between the joint and the tube. Next to this, the tubes present local deformation on the two locations where the load is applied as a result of the pressure. Moreover, until a certain amount of load is reached and the whole system is stiffened small deformations might occur, both within the testing rig but also due to possible imperfections of the tubes.

Until 20mm of vertical displacement the curves mostly follow a linear trend, whereas afterwards and until the breaking point the curves show continuous increase of the resistance. The plastic deformation of the specimen is obvious with wrinkles appearing on the compression side that intensify as the load rises, providing a warning for the upcoming failure. The specimen fails on tension, as shown in figure 4.19.

Table 4.11 'Timber plate puzzle node' | Testing configuration 1 | F_{max} Ultimate (Point A: support, Point B: Middle)

Configuration 1 Specimen Nr.	F_{max} (N) (Ultimate)	Displacement (mm) Point A	Displacement (mm) Point B	Time (sec)
Specimen 1	8098.53	32.68	-0.10	393
Specimen 2	8358.69	36.36	0.29	437
Specimen 3	8625.24	39.03	1.68	469
Specimen 4	8299.58	38.81	1.83	467
Specimen 5	8322.018	35.73	0.50	430
Specimen 6	8363.874	36.05	0.94	433

Graph 4.4 'Configuration 1', average graph line (red color) and linear trend line (black dashed-line). To explain the legend, the initials 'CW' were used as a code name for this case study, for the testing process, and 'c1' indicates the number of 'configuration 1'.

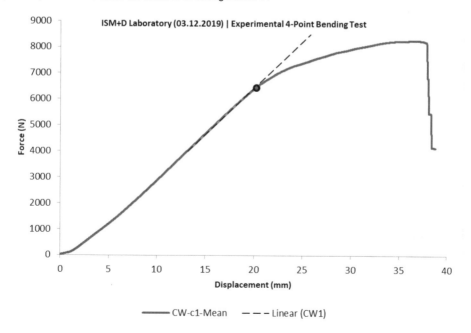

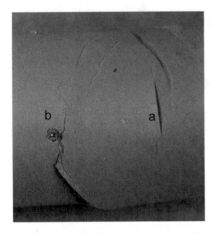

The wrinkle marked with 'a' in fig. 4.20 (top) is a result of the steel element that is used to apply the force, whereas the wrinkles above area 'a' are a result of compression forces.

The crack propagation shows that the screw (b) fixation changed the direction in which it developed.

The crack is most times initiated directly at the bottom side at the area where the force is applied. From there, it develops further through the consecutive layers of the tube. It seems to follow every time the winding line (in fact the layers overlap to provide stability). Then, as the wooden joint is much stronger, the crack develops towards the outer boundaries of the joint.

Smoothening the edges and implementing a ring for compression on the tube, close to the end of the joint, could prevent the failure in this area.

No cracks occurred on the wooden reinforcements.

Figure 4.20 Observation of failed specimen (top) fracture around screw, (middle) failure at the edge of the timber joint, (bottom) displacements inside the tube.

Testing configuration 2

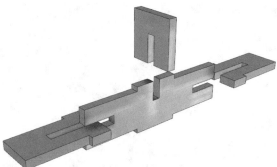

Figure 4.21 'Configuration 2', (top) failed specimens, (bottom) assembly of specimen.

This testing configuration regards the second strongest direction of the node. The continuous plate is the one oriented with its surface on the same plane as the acting forces (fig. 4.21) and it is enough to stiffen the joint sufficiently, until the paper tube fails. The failure modes presented are similar to those described for the first configuration. In total three repetitions were conducted. Figure 4.21 confirms the relation in the failure mode of all three specimens and also with the specimens from series 1.

All specimens were disassembled without any difficulties that could rise due to damaged screws for example, as shown in fig. 4.22, bottom.

Figure 4.22 'Configuration 2', (top) testing rig, (bottom) disassembled failed specimen.

Graph 4.5 'Configuration 2', testing results, force versus displacement curves. To explain the legend, the initials 'CW' were used as a code name for this case study, for the testing process, and 'c2' indicates the number of 'configuration 2'.

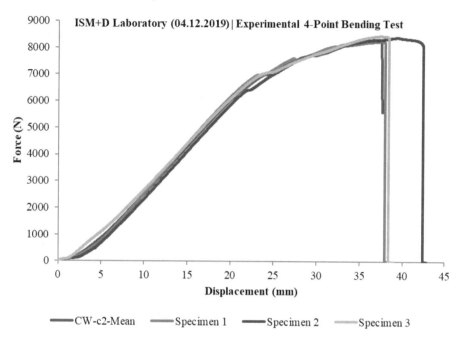

Table 4.12 'Configuration 2', selected values.

Configuration 2 Specimen Nr.	F_{max} (N) (Ultimate)	Displacement (mm) Point A	Displacement (mm) Point B	Time (sec)
Specimen 1	8240	37.33	0.64	449
Specimen 2	8389	39.34	1.50	473
Specimen 3	8445	37.44	-0.01	450

The values for the Mean maximum $F_{Ultimate}$ and F_{Linear}, as well as the respective vertical displacement, are presented in table 4.13. These values are pretty close to the results from the first series of experiments (configuration 1).

For the Mean maximum $F_{Ultimate}$ a gap of 205 N between the higher and the lower value are noted (table 4.13). For the Mean maximum F_{Linear} the same gap rises to 636 N approximately.

The sensor that was touching the center of the middle vertical plate - point B - (fig. 4.22), monitored changes in the position of the joint that lie within the range of 0 to + 2 mm (table 4.12) approximately. Based on the 2.5mm tolerance that exists between the joint and the tubes, this range of displacement is expected.

Graph 4.6 'Configuration 2', average curve. To explain the legend, the initials 'CW' were used as a code name for this case study, for the testing process, and 'c2' indicates the number of 'configuration 2'.

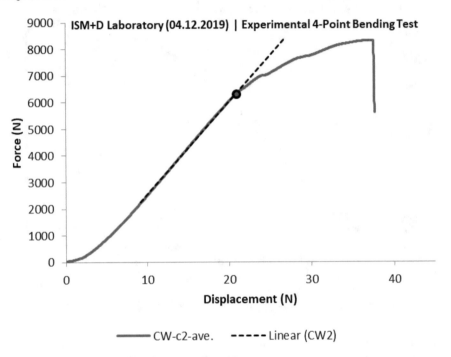

Table 4.13 'Configuration 2', selected mean values.

Configuration 2, Mean values	Mean F_{max} Ultimate (N)	Displacement Point A (mm)	Mean F_{max} Linear (N)	Displacement Point A (mm)
	8307	37.10	6300	20.8

Testing configuration 3

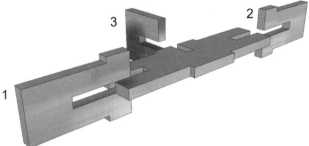

Figure 4.23 (Top) 'Configuration 3', testing rig, (bottom) Assembly of specimen

This is the weakest direction of the node. Three repetitions were conducted. The results indicate indeed significantly lower performances comparing to the previous two series of tests. All specimens failed in the same way. The stresses caused by bending concentrate in the horizontal plate, which finally cracks, in the middle area of the joint. The vertical plates 1 and 2 are not in perfect contact with the middle vertical plate 3. As mentioned before (section 'production and prototyping'), the small voids, between the plates, are a result of the curved inner corners. Even if this contact would be perfect, it is estimated that the results would still be in the same range of values that are presented in table 4.14.

Table 4.14 'Configuration 3', selected mean values

'Configuration 3', Mean values	Mean F$_{max}$ Ultimate (N)	Displacement Point A (mm)	Mean F$_{max}$ Linear (N)	Displacement Point A (mm)
	5104.50	27.85	4300	20.56

In tables 4.14 and 4.15 the resulting values from the tests performed are shown. The Mean maximum F$_{Linear}$ is 4566 N, with a gap of approximately 2 kN rising between the highest and the lowest values. This gap indicates that further tests are certainly required. It is suspected though that in the case of the specimen characterized by the lowest value, material imperfections are responsible for this condition, as it is known that within plywood often voids exist.

The curves in graph 4.7 show that after the specimens reach the maximum Force, immediate loss of strength occurs and they fail almost abruptly. As a consequence, even though the joint shows warning signs for the upcoming failure (small cracks), it is obvious that 'configuration 3' presents important weaknesses that should be considered in the design process.

Graph 4.7 'Configuration 3', testing results, force versus displacement curves. To explain the legend, the initials 'CW' were used as a code name for this case study, for the testing process, and 'c3' indicates the number of 'configuration 3'.

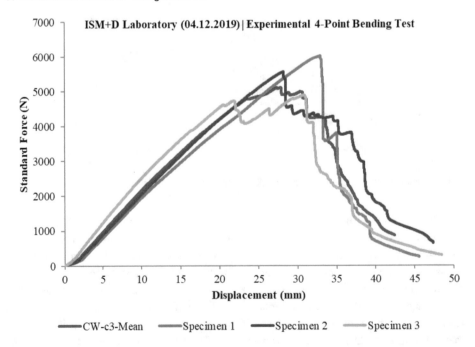

Table 4.15 'Configuration 3', selected values

Configuration 3 Specimen Nr.	F_{max} (N) (Ultimate)	Displacement (mm) Point A	Displacement (mm) Point B	Time (sec)
Specimen 1	6026	33	-6.50	395
Specimen 2	5563	28	-5.63	339
Specimen 3	4892	31	-9.56	370

Graph 4.8 'Configuration 3', average graph line. To explain the legend, the initials 'CW' were used as a code name for this case study, for the testing process, and the number is indicative for 'configuration 3'.

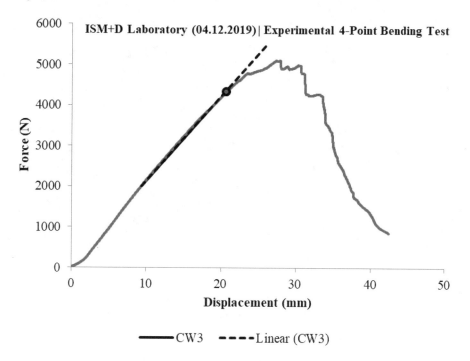

Figure 4.24 'Configuration 3', side views of the tested specimen.

In figure 4.24 (top) the failure mechanism is shown. The different veneers of the plywood crack and split one by one, until, in continuation, the plate cracks open.

The wooden reinforcements are also laminated on both sides of the vertical middle plate 3. However, due to the orientation of the fibers, the bond is not very strong. Still, during the bending of the system, as tensile forces occur in that area, delamination is caused within the middle vertical plate, whereas the glue bond on its side sustains (fig. 4.24, bottom and 4.25, top).

As a result of the significant vertical displacement, the screws also appeared to have suffered higher stresses than in the previous experiments and few fell out when the specimen failed. Nevertheless, all specimens were disassembled without significant issues (fig. 4.25, bottom).

Figure 4.25 'Configuration 3' (top) cracked joint, (bottom) dissassembled failed specimens.

Testing the limits of the screws

Figure 4.26 Testing process

The force is applied on the paper-tube, whereas the timber joint ends about 5cm lower. Imperfections in the set-up are observed at the contact area between the steel plates and the specimen, which are not particularly significant for the testing results. These are located specifically on the upper side, along the surface of the paper-tube and more significantly at the bottom side, along the timber cross. In the cross internal tolerances lead to lack of contact between the one timber plate and the base steel plate. This condition increases the concentration of stresses on the other timber plate and the center of the timber cross. Still, no damage was observed on the timber plates.

Graph 4.9 Pressure-test, force versus displacement curves.

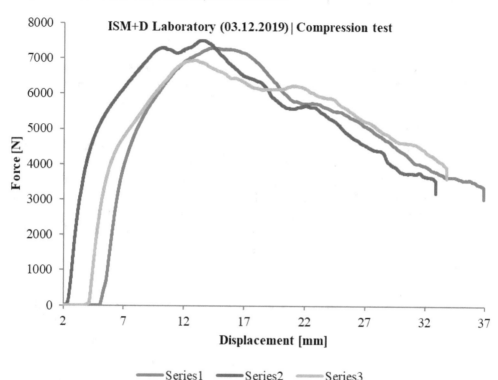

The tolerances mentioned above are responsible for the high displacement while the force is still 0, as shown in graph 4.9.

As expected, for all three specimens, the cause of failure was damage on the side of the paper tube and consequently the screws. The screws teared the paper and in continuation the vertical displacement caused the screws to bend. The rest of the specimen remained intact.

Following graph 4.9, the Mean Maximum Force carried, within the linear part of the curves, is approximately 3,5 kN, with a respective vertical displacement about 5mm. On the other hand, the Ultimate Maximum Force is up to 6kN, with a respective vertical displacement about 10mm.

According to the testing results, a shearing load equal to 44kg per screw is calculated, before plastic deformation occurs in the system.

In the actual joint, the timber plates are designed to support the ends of the tubes, so that the shear forces acting on the screws are minimized.

Local FEM Analysis

Multi-axial node

Hereby essential input for the build-up of the ANSYS model is reported. In continuation, the output is discussed. Selected exports provide a good impression of the output from the analysis.

Material Properties:

- Birch Plywood: young's modulus $E_m \perp$ 7462 MPa and characteristic strength $f_m \perp$ 34.3 MPa, both retrieved from table 3-2, page 19 (Finnish-Forest-Industries-Federation-Manufacturers, 2002) and Poisson's ratio 0.426 retrieved from table 4-2, page 4-3 of (David W. Green, 2010)
- Douglasie natural timber: Young's modulus $E_{0,mean}$ ‖ 7000 MPa, characteristic strength for bending $f_{m,k}$ 14 MPa (Eurocode5, 2009), Poisson's ratio 0.292, table 4-2, column μ_{LR}, page 4-3 (David W. Green, 2010)

Contact properties:

- Between the plywood plates: frictional (0.4)
- Between the plywood plates and the reinforcement (wooden profiles): bonded
- Between the wooden parts of the joint and the paper-tube: frictional (0.4)
- Between the steel elements and the wooden joint or the paper-tubes: bonded

The forces and moments applied on the system are according to table 3.11, paragraph 3.3.2. The strongest axis of the joint is oriented in the direction with the highest forces and moments, which is this of members 9 and 10, following also the values stated in the same table. Figure 4.27 (bottom) shows indeed the higher concentration of stresses along this axis, especially where there is contact with the tubes, even though the ultimate values are expressed within a low range. Overall, the 'Maximum Principal Stress Analysis' shows that the most critical areas lie in the surroundings of the mechanical fixations. The peak stresses occur on the screws. As the loads are applied on the paper tubes, the screws are the only elements that intersect with both these and the joint. In continuation, the wooden reinforcements that are laminated on the joint and screwed together with the tubes suffer the majority of stresses. In principle, the stresses are within the capacity of the material but close to the screws the conditions become critical and thus failures (cracks) could be expected. Still, the stresses occurring on the tubes exceed the capacity of the component. The structural system suffers compression on the top side and tension at the bottom, as it can be seen in figure 4.27 (bottom). Moreover, as expected, high stresses occur at the bottom corners, close to the end of the tubes. These observations indicate that further optimization of the joint is required. These may involve the use of stronger paper-tubes, stronger wood for the reinforcements and combination of screws with pressure profiles, to optimize the distribution of stresses etc.

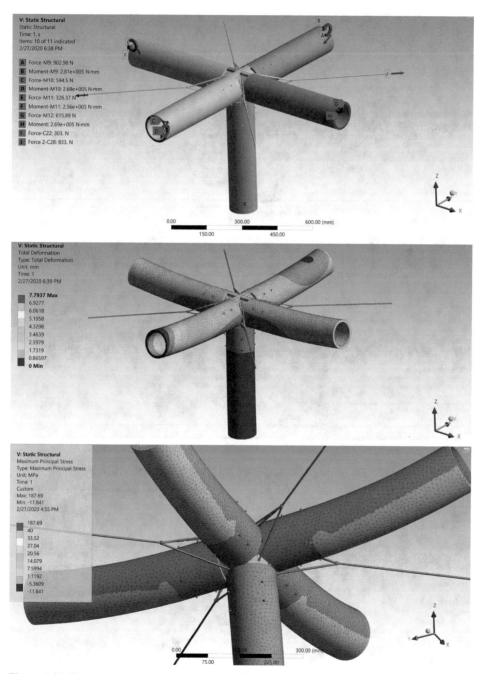

Figure 4.27 (Top) Loads, (middle) total deformation analysis, (bottom) maximum principle stress analysis.

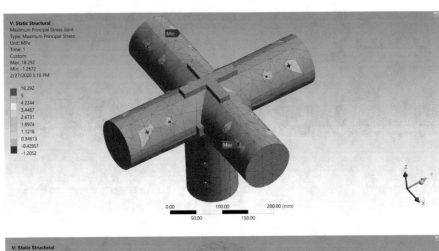

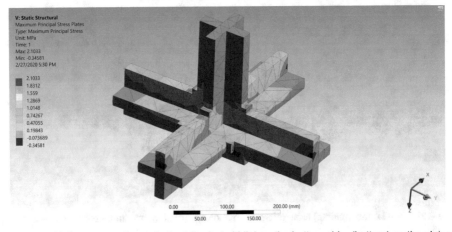

Figure 4.28 Stresses on the node (top) Peak, (middle) on the bottom side, (bottom) on the plates.

Analysis of Linear Configurations 1, 2 and 3

The forces applied are according to the $F_{max\ Linear}$, as defined by the results from structural testing. The contact properties follow the boundary conditions that apply for the analysis of the multi-axial joint. In the following pages, for each of the three studies, selected exports are presented that show the boundary conditions and results from the FEM analysis. Below key-observations are reported.

- 'Configuration 1'

Overall, peak stresses occur at the areas around the screws. The range of stresses can be better evaluated for the individual components, as shown in fig. 4.30. As indicated by the values identified there, the reinforcements could be damaged where screws are anchored, whereas the timber plates suffer high concentration of stresses at the bottom corners of contact with the tubes. Moreover, the tubes suffer particularly high stresses, comparing to the capacity of the material, at the bottom area where the joint ends. Thus, the tubes are expected to fail at that area, as it also happened at the actual testing process. The maximum stresses occurring at the fasteners exceed the capacity of the selected screws (S235). The stresses occurring at the supports are comparable to the values calculated based on the results from structural testing (fig. 4.29 bottom). Furthermore, the total deformation calculated by the software is slightly smaller (by 4mm approximately) comparing to the structural test, a sensible range considering that imperfections that apply in real conditions don't exist in this simplified idealized system.

- 'Configuration 2'

The observations are similar to the previous case. A further observation is related to the plate placed in the middle of the assembly. The contact areas between all adjacent edges and surfaces present high concentration of stresses, as it can be seen in fig. 4.32, middle.

- 'Configuration 3'

Based on the current analysis three failure modes can be detected, cracking of the middle horizontal plate, failure at the corners of the reinforcements, where they meet the plates, and failure of the system at the ends of the joint. At first, in the middle of the horizontal plate (fig. 4.34, middle) the matrix of stresses indicates the pattern of tension and compression within its thickness that makes the plate bend and eventually break. If plywood was modelled as an orthotropic material the results would be closer to these from structural testing. Secondly, the stresses at the inner corners, at the reinforcement (4.34, top) present a critical condition. Then, high concentration of stresses occurs in the contact areas between the cross and the tubes, as it can be seen in fig. 4.34, middle and bottom. There the highest deformations are calculated. As the joint is not strong enough in the middle, the stresses concentrate on the sides, as shown in the graphs of fig. 4.33. Following this analysis, the system is supposed to fail at the bottom side close to the edges of the joint. Moreover, the tube suffers deformation that is caused by tension at the areas of the screws and compression at the contact with plates.

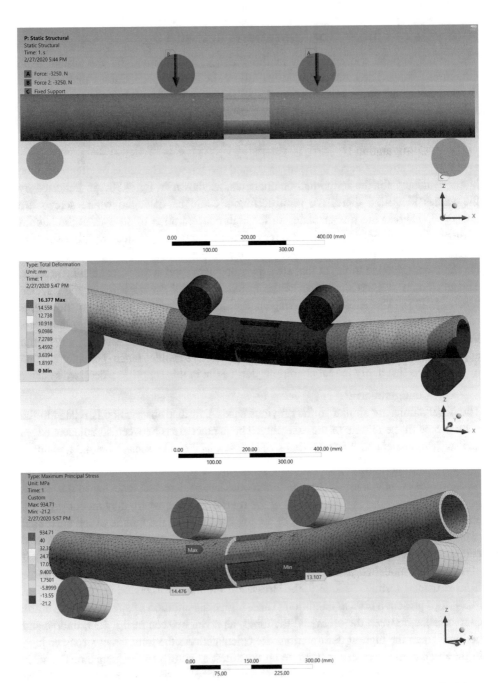

Figure 4.29 'Configuration 1', (top) Loads, (middle) Total deformation, (bottom) Maximum principle stresses.

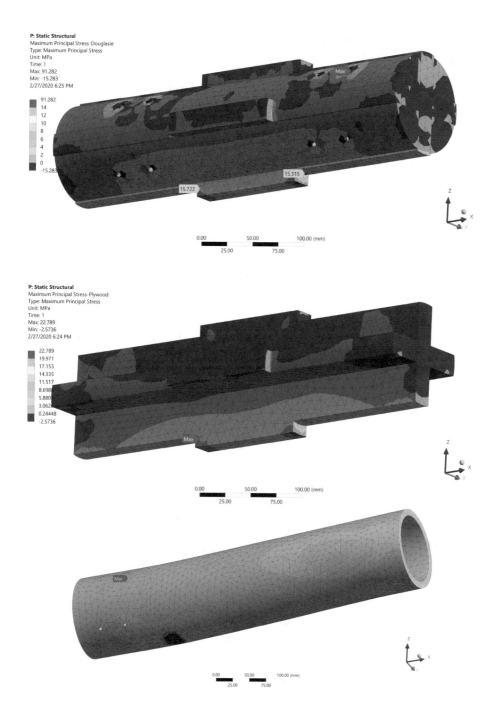

Figure 4.30 'Configuration 1', Distribution of stresses on the individual components.

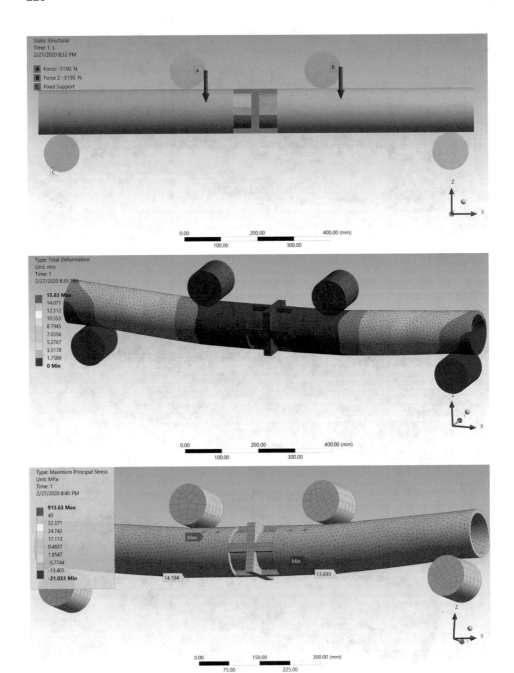

Figure 4.31 'Configuration 2', (top) loads, (middle) total deformation, (bottom) maximum principle stresses.

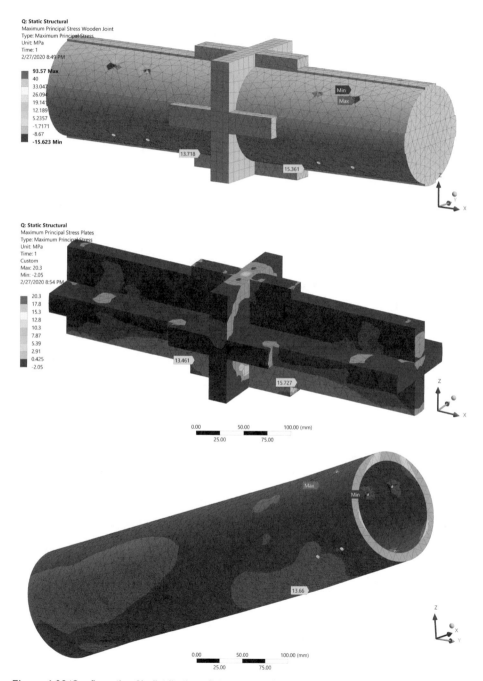

Figure 4.32 'Configuration 2', distribution of stresses on the individual components.

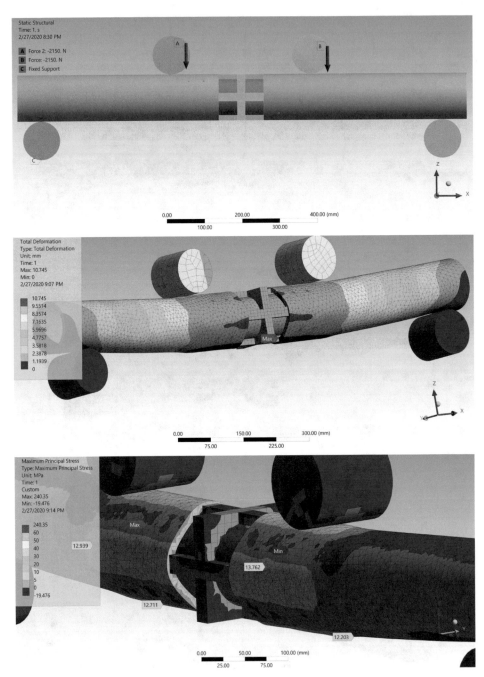

Figure 4.33 'Configuration 3', (top) Loads, (middle) Total deformation, (bottom) Maximum principle stresses.

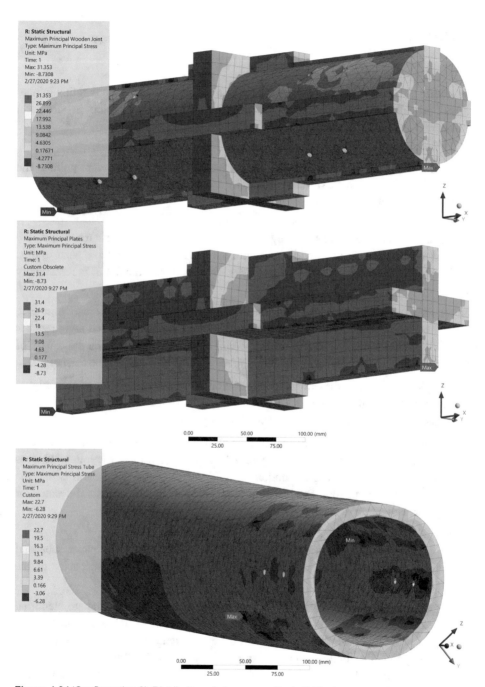

Figure 4.34 'Configuration 3', Distribution of stresses on the individual components.

Summary

- Design and assembly aspects

The studies about 'design alternatives' and 'assembly methods' indicate that this joining method provides a wide spectrum of possibilities for implementation in structural systems of different geometries.

- Production and prototyping process

The production methods are fairly simple with minor issues. An important point is the post-processing of the plates after cutting, by treating the inner corners, to achieve efficient form-fitting.

- Global observations from structural testing

The testing process confirms the hypothesis that 'configuration 1' represents the strongest axis for a multi-axial node built with the current joining method. Next to this, 'configuration 2' presents similar performances whereas 'configuration 3' is remarkably weaker.

For all three configurations that are tested, the margin between max and min F_{max} 'linear' and 'ultimate' is significant. Especially for 'configuration 3' this margin reaches the value of 2 kN. Thereby, further testing would be needed to determine the causes of this margin and also identify more securely the maximum force that can be carried.

In principle, $F_{max,linear}$ in directions 'configuration 1' and 'configuration 2' is higher comparing to this of the the paper tube (see paragraph 4.2, table 4.7). However, in the direction of 'configuration 3' $F_{max,linear}$ is significantly lower.

As explained before, the structural system fails at the area in which it suddenly becomes weaker. For configurations 1 and 2 this area is located at the ends of the joints, whereas for the third configuration it is the middle of the joint. A global conclusion is that the joint does not contribute significantly as a reinforcement for the structural system for bending, especially due to contact issues.

- Global observations from 'ANSYS FEM Analysis'

Based on the idealized conditions assumed for the modelling of the paper tube and observing the resulting stresses, it is expected that these exceed the capacity of the component.

Besides the tubes, the areas where the capacity of the material is exceeded are identified on the reinforcements. This is a result of several parameters. Some are related to the build-up of the ANSYS model, whereas to a different extent it is a design problem. To explain this point, the limited contact between the joints and the inner surface of the paper-tubes increases the role of the screws. This way, the forces are transferred through points or lines instead of surfaces.

In the multi-axial joint examined in 'ANSYS' the design of 'configuration 3' is integrated in the lower part, towards the column-support (see fig. 4.8) and therefore does not suffer significant bending, as in the structural testing experiments.

Comparison between the results from structural testing and the 'ANSYS analysis', shows that the resulting maximum stress at the area of the middle supports (point of force application) is in the same range.

Graph 4.10 Collective data for the case study 'Timber plate puzzle node'. Comparing the mean curves for all three configurations tested and also paper tubes. To explain the legend, the initials 'CW' were used as a code name for this case study, for the testing process, and the numbers 1, 2, 3 indicate the respective configuration.

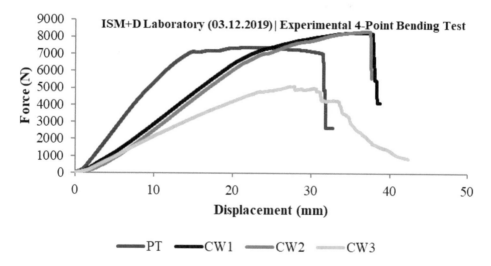

Table 4.16 Overview of results from 4-point bending test (mean values).

Timber plate puzzle node	Mean F_{max} Ultimate (N)	Displacement Point A (mm)	Mean F_{max} Linear (N)	Displacement Point A (mm)
Configuration 1	8253	36.7	6500	20.3
Configuration 2	8307	37.1	6300	20.8
Configuration 3	5104.50	27.85	4300	20.56

Literature

David W. Green, J. E. W., David E. Kretschmann. (2010). Mechanical Properties of
    Wood. Retrieved from https://www.fs.usda.gov/treesearch/pubs/37427
Eurocode5. (2009). Strength and stiffness properties and density values for structural
    timber strength classes, (in accordance with Table 1, of BS EN 338: 2009).
    Retrieved from https://download.e-bookshelf.de/download/0003/7612/73/L-X-
    0003761273-0007909753.XHTML/index.xhtml
Finnish-Forest-Industries-Federation-Manufacturers. (2002). *Handbook of finnish
    plywood*.
McQuaid, M. (2003). *Shigeru Ban*. London N1 9PA: Phaidon Press Limited, Regent's
    Wharf All Saint's Street.

4.2.1.2 Timber block node

Figure 4.35 Prototype of triaxial joint, production: 07.2019.

This joining method provides increased contact surface between the structural members and also advantages for the integration of mechanical fixations. These are some of the benefits comparing to case study 4.2.1.1. Thus, it has been used in more durable structures. Examples of application in construction projects are referenced in paragraph 2.1.1, table 2.1, project numbers: 3, 4, 12, 20. The main typologies of assemblies are visualized in paragraph 2.2.4.1, 'Massive wooden nodes', in figures 2.33 – 2.36.

Hereby, the main idea is to simplify the matrix of the individual components and develop a multi-axial joint that relies on single wooden elements made of natural wood that are connected in terms of form-locking and secured with mechanical fixations. This is the main difference comparing to the reviewed examples. To achieve this, a critical aspect is the manufacturing process that demands either advanced skills in wood-working or more advanced technology for automated production. Next to this, the structural performance in the different axes is the main challenge, as explained later in this chapter. An example of a bending test for such a joint is presented in literature (McQuaid, 2003, p. 79), where the failure appears on the paper-tube, a continuous crack that follows the bolted fixations.

Table 4.17 Characteristic joining method – 'Timber block node'.

Form-locking			Timber blocks
Force-closure	**Friction**		Node-tubes
	Pressure		Bolted node
	Tension		
	Inertial Forces		Tensile cables
Material-closure			Corner reinforcements glued on timber node

Detailing alternatives

The most important aspects for the structural integrity of the node are optimum contact between the surfaces of the wooden parts - form of plug-in elements -, as well as a strong connection between those - mechanical or other fixation -.

There are two main design alternatives. The first one is to design an entirely hidden plug (see figure 4.38) - that is more minimal for aesthetics. However, in this case, errors in production can easily worsen structural performance, as indicated by the testing results of configuration 2 (see page 237). Moreover, a steel thread is integrated in the core of the joining system, to hold the elements under pressure, improving thus the structural performance and also acting as a measure for security in case of structural failure.

The second option is to design an extended outer base for the insert (see figures 4.35 - 4.37 and also preliminary testing series that are presented in paragraph 6.2 appx.), to provide better support against tilting and thus enhance the stability of the assembly. Such a design approach leads to a connection with a bigger size that from an aesthetical perspective could be viewed as a disadvantage. Still, on the positive side, it creates better conditions for the implementation of mechanical fixations (pressure plates) between the wooden parts, as shown in fig. 4.37.

In both cases, potential use of adhesives to absorb tolerances between the wooden blocks is possible. Additionally, the integration of a steel thread for pressure of the wooden blocks is advisable in both cases.

A different approach for the design of the joint is presented in the reference project of a temporary 'office studio' of S. Ban, in Paris (Miyake, 2009, p. 95), also shown in chapter 2, page 55 (fig. 2.36). The joint is divided in two symmetrical sides that are joined exactly in the middle, with biscuits and are secured from the side with steel plates. This option might present advantages for the production process, but has downsides on the stability, as the center of a node is a particularly critical area.

Assembly techniques

The recommended assembly method is clarified in table 4.17 and examples are demonstrated in figures 4.37 and 4.38. The tolerance between the joint and the diameter of the paper tube is important both for the assembly process and the stiffness of the system. When tolerances are approximately 0, then high pressure is required to achieve the fitting. Like in the previous case study, a slightly higher tolerance close to the edges of the multi-axial joint can ease the assembly process. Regarding the mechanical fixations, in principle, bolted joints are less sensitive, comparing to the screwed ones. Hence, the bolted assemblies are easier to handle (less likely to damage).

Imagining the assembly of an entire structure (rectangular or with a slanted or pitched roof), a possible concept is to assemble the entire roof, together with the eave joints, so that on the site, the assembly can be realized in three steps. Placement of the fundament first, the columns secondly and the entire prefabricated roof in the end. In the case of an arch, all sections could be prefab and then connected on the site.

Relocation of the structure is in principle possible. Depending on the time-frame and the boundary conditions, the paper tubes might need to be replaced. Next to this, implementation of slightly bigger screws is recommended.

Production and prototyping

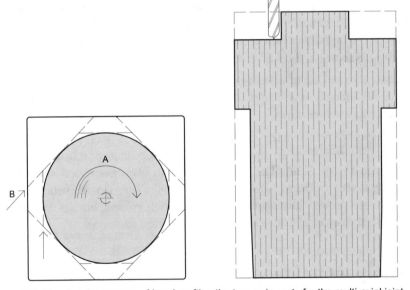

Figure 4.36 Planning the process of hand-crafting the two main parts for the multi-axial joint.

In this case study, the wooden inserts are hand-crafted. For the long cylindrical shaped inserts, at first the form was approximated with angular cuts and then the surfaces were smoothened with basic wood-working tools and methods. For the short insert, designed to fit the wooden parts together, free-hand routing was performed. An alternative method for automated production is 5D CNC milling that with modern machines can even be performed around cylindrical surfaces. For the mechanical fixations long drill-bits

(25 cm) were used and custom-made guides were built to achieve straight drilling along the parts of the joint, for the integration of the steel thread (M8 diameter was selected). To optimized the distribution of stresses at the ends wide steel plates instead of common rings may be used.

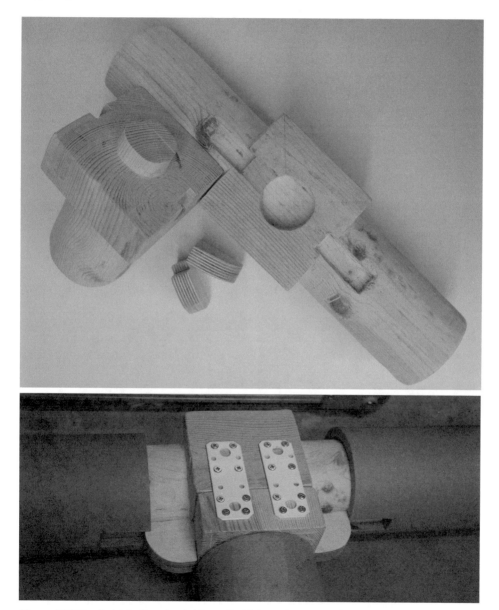

Figure 4.37 Plug-in joint with extended base (top) joinng elements and (bottom) assembly method.

Figure 4.38 Components and assembly of specimen with minimal insert (final series of testing).

Structural Testing

Testing configuration 1

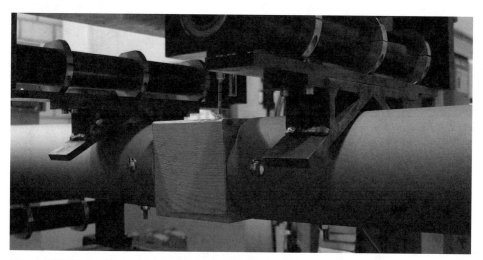

Figure 4.39 'Timber block node', 'configuration 1', specimen 2.

In total three specimens were tested. The failure mode is systematically observed on the paper tube and basically regards bursting of the paper-tube along the winding line.

As it can be observed in graph 4.11, for the first few millimetres of vertical displacement the system is moving until certain tolerances are absorbed and then it gains stability.

Hence, there is an uncertainty factor due to imperfections in production, especially around the perimeter of the side cylinders and the contact area between edge of tube and flat end of the cubical part of the joint.

The Mean $F_{max,Ultimate}$ is about 8071 N, followed by a vertical displacement of 38.45 mm (see table 4.19), whereas the margin between maximum and minimum $F_{max,Ultimate}$ is about 607 N (see table 4.18). The margin between the maximum vertical displacement for the same group of values is about 11,69 mm.

The $F_{max,Linear}$ is significantly lower, about 6500 N (see graph 4.12 and table 4.18).

The vertical displacement monitored by the sensor in the middle of the set-up is up to 3,4 mm and can be explained by the tolerances between the joint and the tube, in combination with the fact that the joint itself is stiff enough to present 0 deformation and instead develop vertical movement as a whole.

Minor issues are observed at the areas of the bolts (fig. 4.40 bottom).

Graph 4.11 Testing results for the 'Timber block node', configuration 1: force versus displacement curves. To explain the legend, the initials 'MW' were used as a code name for this case study, for the testing process, 'c1' indicates the number of the configuration and the numbers 1,2 and 3 the different specimens.

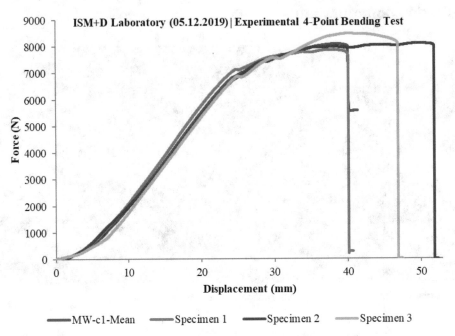

Table 4.18 'Timber block node' | Testing configuration 1 | F_{max} Ultimate (Point A: support, Point B: middle).

Configuration 1 Specimen Nr.	F_{max} (N) (Ultimate)	Displacement (mm) Point A	Displacement (mm) Point B	Time (sec)
Specimen 1	7858	38.04	1.82	457
Specimen 2	8098	49.73	3.38	598
Specimen 3	8465	40.28	2.56	484

Graph 4.12 'Timber block node', configuration 1, force versus displacement average curve. To explain the legend, the initials 'MW' were used as a code name for this case study, for the testing process and 'c1' indicates the number of the configuration tested.

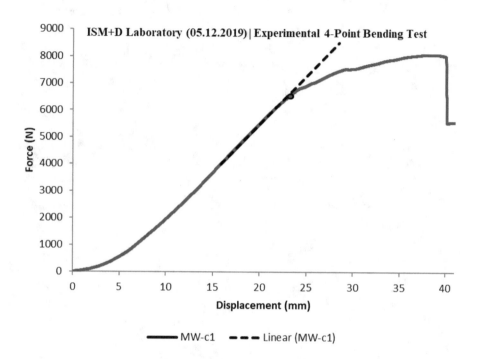

Table 4.19 'Timber block node', configuration 1, selected mean values.

Configuration 1, Mean values	Mean F_{max} Ultimate (N)	Displacement Point A (mm)	Mean F_{max} Linear-part (N)	Displacement Point A (mm)
	8071	38.45	6500	23

Figure 4.40 (Top) testing process (spec. 3), (middle) failed specimen (spec. 2), (bottom) minor impact around bolt.

Testing configuration 2

Figure 4.41 'Timber block node', configuration 2, specimen 1.

Three specimens were tested. Here the uncertainty factor due to imperfections of the sub-elements of the assembly is higher, as the joint is not continuous. This is understood also by the testing results. The Mean $F_{max,Ultimate}$ is 5305.5 N, followed by a vertical displacement of about 60 mm. The margin between the maximum and minimum $F_{max,Ultimate}$ is about 603.4 N. For the same group of values, the margin between the vertical displacement is about 22mm. The Mean $F_{max,Linear}$ is about 3500 N followed by a vertical displacement of 22 mm. The sensor measuring the vertical displacement in the middle of the set-up monitored a displacement of about -17 mm in two of the repetitions, as shown in table 4.20. This is the only test in which it is significant and in the opposite direction (steel thread reaction). Furthermore, the functionality of the joint is problematic, as indicated by the high vertical displacement. The joining system bent significantly, due to issues in the contact between the wooden parts (see figure 4.44). Thus, optimization in the design and fabrication is required.

Figure 4.42 Testing configuraion 2, failed specimen.

Graph 4.13 Testing results for the 'Timber block node', configuration 2: force versus displacement curves. To explain the legend, the initials 'MW' were used as a code name for this case study, for the testing process, 'c2' indicates the number of the configuration and the numbers 1,2 and 3 the different specimens.

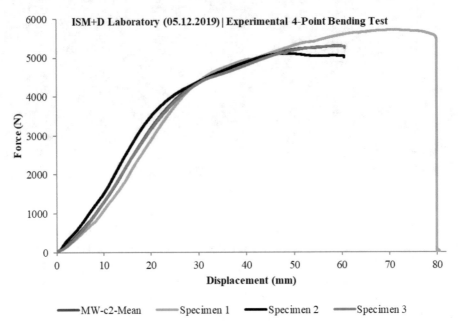

Table 4.20 'Timber block node' | Testing configuration 2 | Fmax Ultimate (Point A: support, Point B: middle).

Configuration 2 Specimen Nr.	F_{max} (N) (Ultimate)	Displacement (mm) Point A	Displacement (mm) Point B	Time (sec)
Specimen 1	5703.4	70.18	-17.63	843
Specimen 2	5100	48.22	0	580
Specimen 3	5286	59.67	-16.80	717

Graph 4.14 'Timber block node', configuration 2, force versus displacement average curve.

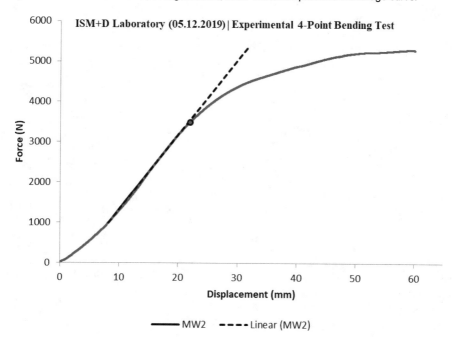

Table 4.21 ‚Timber block node', configuration 2, selected mean values.

Configuration 2, Mean values	Mean F_{max} Ultimate (N)	Displacement Point A (mm)	Mean F_{max} Linear (N)	Displacement Point A (mm)
	5305.5	59.76	3500	22

Figure 4.43 (Top) testing process, response to loading (specimen 1), (bottom) impact of bending on bolted end.

Figure 4.44 Configuration 2, the plastic deformation visible on the three tested specimens.

Testing the bolted fixations

Figure 4.45 Experimental test, to observe the failure mechanism of the bolted joints under shear.

The failure mode is as expected (see fig. 4.45). A point was to get a glimpse on the limit of the F_{max} comparing to case study 'Timber plate puzzle node'. Even though only a single repetition was performed, it is obvious that within the limits of a linear trendline, the performance of this bolted joint is much higher. For the axis examined in configuration 2 it is expected that a similar test would also indicate an advantage of the current joint for shearing, as the screws are bigger.

Graph 4.15 Pressure test, force versus displacement curve.

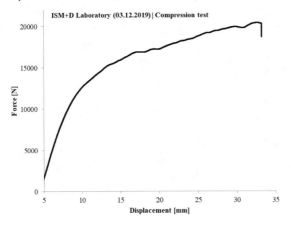

Local FEM Analysis

Multi-axial node

The aim is to analyse the performance of the node when subjected to the load combination described in paragraph 3.3.2, table 3.11. At first, the input used to build-up the 'ANSYS model' is provided. The geometry of the joint is presented in fig. 4.46 (top). The design of the bolts is simplified, to reduce the complexity of the mesh. The material properties assumed are the following:

- Natural Timber: Young's modulus 11000 MPa (Eurocode 5, C24), (KVH), maximum stress 24 MPa and Poisson's ratio 0.38 (David W. Green, 2010)
- Birch Plywood: same as in case study 'Timber plate puzzle node', page 214.

The contact properties assumed are as following:

- The elements of the wooden node are connected with each other in terms of friction. Between the wooden node and the paper tubes also frictional contact applies. In both cases the friction value is 0.4.
- The contact between the steel elements and the joint or the tubes is considered as bonded.

The paper tubes are expected to fail first, as the resulting stresses are high, comparing to the capacity of the material, also considering the simplification of the material model. As shown in fig. 4.46, the paper tubes are subjected to tension along the bottom side and compression along the top side. The critical failure is expected to occur on the side of tension. The plug-in joint defines the boundary for the zone within which tension occurs. Then, mechanical fixations (either screws or bolts) placed in the same area become critical points (example shown also in figure 4.46 bottom). This way, the conditions described beforehand create a good impression about the failure of the system, together with the results from structural testing. In a different way, due to the idealized geometrical conditions that apply for the ANSYS model, the contact area between the wooden blocks in configuration 2 does not express the same problems, as in the testing process (fig. 4.43 -top left). Figures 4.46 – 4.48 help to understand the performance of the wooden node. The most critical areas identified appear along axis y of the model, in the area where the min value occurs, as already mentioned. In fig. 4.46 (bottom) the problem within the contact area between the wooden blocks and the steel thread is visible, with high tension in the middle and pressure at the bottom, following the direction of the forces and the expected reaction and deformation of the system. In direction x higher stresses occur at the contact area between the central block of the node and the tubes, as significant normal forces are transferred. Still the values are within a good range. As already mentioned, high stresses occur at the areas where joint and tubes are connected with fasteners. More specifically, the 'Maximum Principal Stress' analysis indicates that for the wooden joint the critical areas differ for configurations 1 and 2.

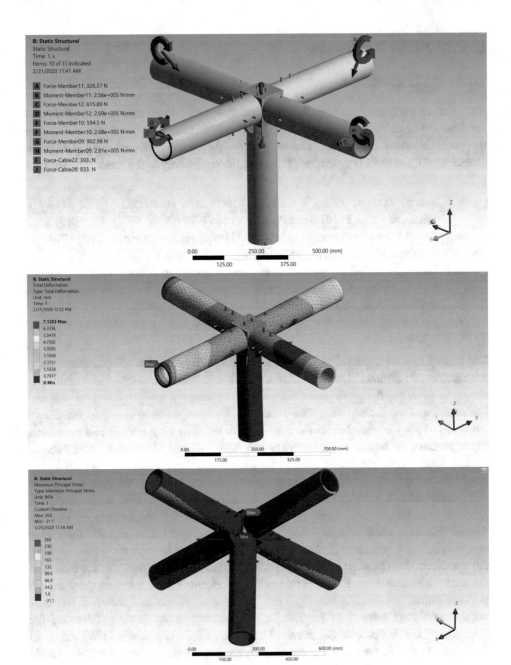

Figure 4.46 (Top) the forces and moments applied on the structural system, (middle) "Maximum Principle Stress" analysis (Solution), (bottom) 'Total Deformation' analysis.

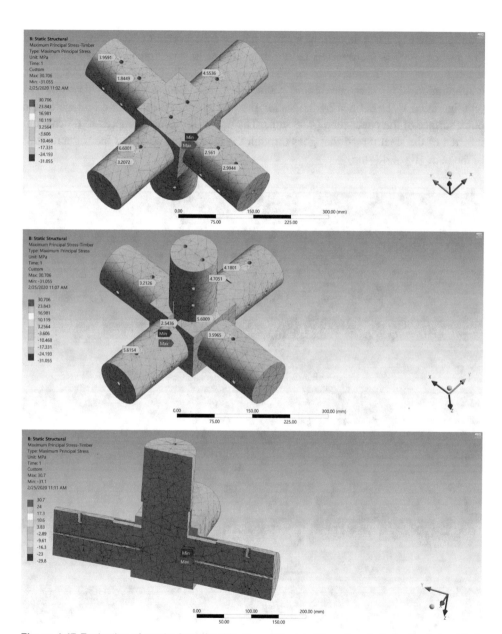

Figure 4.47 Evaluation of results from the "Maximum Principle Stress" analysis for the timber block node, (top) top view, (middle) bottom view, (bottom) section that reveals the high stresses occurring at the contact areas - junctions- between the timber blocks and the steel thread.

In direction x, following the co-ordinate system in fig. 4.47, cracks could be initiated at the areas of the holes for the bolts, from the surface towards the inner core of the wood (configuration 1). Even though the stresses mostly do not seem to exceed the capacity of the material, this is a suspected failure mechanism. In direction y, the critical area is located along the middle axis of the joint (fig. 4.47 -bottom- and 4.48 – top, configuration 2). The maximum value -tension- occurs in the direction of 'configuration 1', as defined by the testing process, close to the start of a bolt, so on the outer surface of the paper-tube (4.48 -top). The minimum value -compression- occurs in the area of the steel thread that goes through the joint in the so-called 'configuration 2' of the node (fig. 4.48 -top).

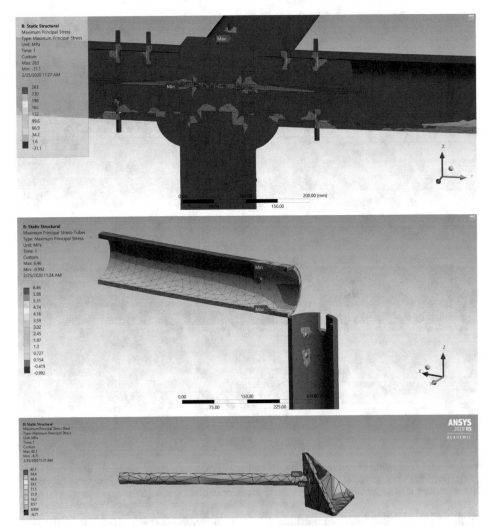

Figure 4.48 'Maximum principle stresses', (top) section along the whole assembly, (middle) section view on paper-tube members, (bottom) cable (28).

Linear configurations

- Configuration 1

The contact properties are the same as these in the analysis of the multi-axial node.

- The critical areas are the same as mentioned in the analysis of the multi-axial node.

- The load magnitude is different, significantly higher and so higher stresses occur in the wooden node.

- Excessive concentration of stresses occurs close to the boundaries of holes around the bolts. To evaluate the situation though, the values selected are outside these locations, as indicated in fig. 4.50.

- The maximum stresses occurring at the support are comparable to the value calculated based on the testing results (evaluation table 4.37).

- Failure is identified on the side of the paper tube, following the stresses simulated for that element (fig. 4.50 -bottom) and the expected capacity of the material, taking into account the aspects related to the material behaviour that have already been stated.

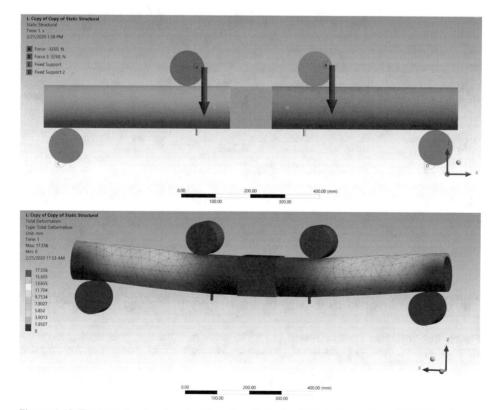

Figure 4.49 (Top) structural system (configuration 1), (bottom) total deformation – 'ANSYS' software.

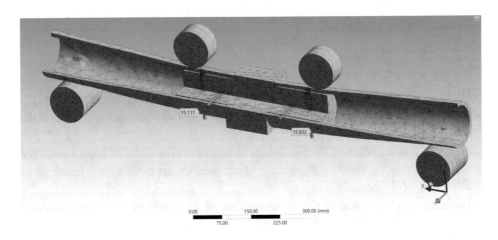

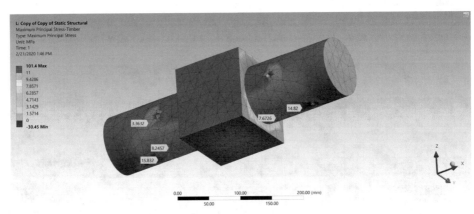

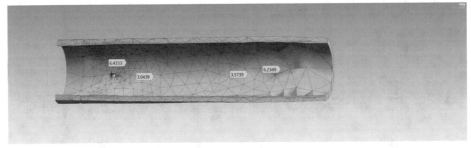

Figure 4.50 'Maximum Principal Stress' in ANSYS software – solution, (top) section along the structural system, (middle) evaluation of stresses on timber-block joint, (bottom) section on paper tube (z-axis perspective).

- Configuration 2

The contact properties are the same as in these in the analysis of the multi-axial node.

- The critical areas are the same as mentioned in the analysis of the multi-axial node and in configuration 1.

- The maximum stresses occurring at the support are comparable to the value calculated based on the testing results (evaluation table 4.37).

- Also, in this case, the failure is identified on the side of the paper tube, in contradiction with the outcomes from structural testing.

- As in all models, the geometrical conditions are idealized. In this particular case, comparing this impression with the testing process, someone can see, based on the total deformation with the current load applied, how much stiffer the joining system could be, if the fitting between the wooden elements would be optimal.

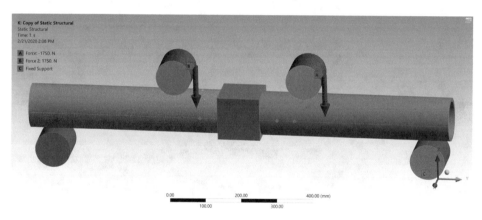

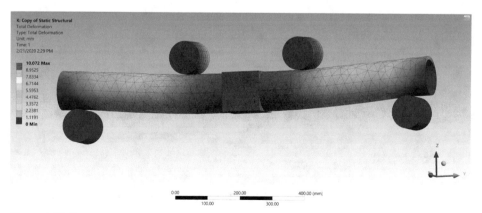

Figure 4.51 (Top) structural system examined (configuration 2), (bottom) total deformation – 'ANSYS' software.

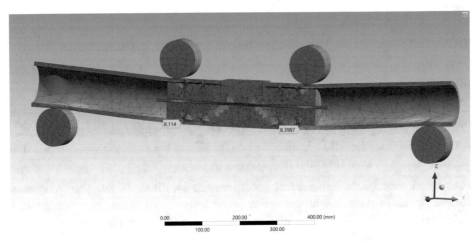

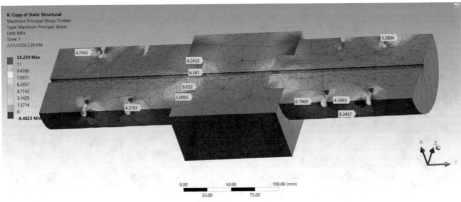

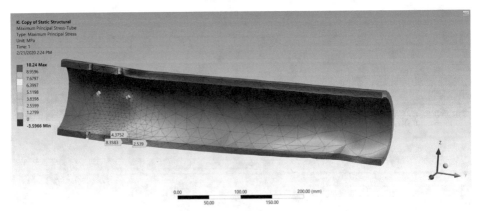

Figure 4.52 'Maximum Principal Stress' in ANSYS software – solution, (top) section along the structural system, (middle) evaluation of stresses on timber-block joint, (bottom) section view along paper tube (x axis).

Summary

- Design and assembly aspects

As detailing alternatives two approaches are suggested for the design of the form-locking joint between the wooden blocks, a fully hidden lock or one with a steadier base. Both can be integrated in different global set-ups and also allow for further detailing of the visible edges of the joint, to serve the aesthetical criteria.

The assembly techniques are simple. When errors in the production are minimized, then the structural system is safer and easier to transport and handle.

- Production and prototyping process

Overall, the prototyping process was successful. An important issue to overcome when hand-crafting is precision. When suitable CNC Milling processes would be available, optimal fabrication would be simple to achieve and the structural performance, also in the configuration 2, would be significantly improved.

- Global observations from structural testing

Regarding the F_{max} values, the margins between the results of each series are significant, as concluded based on tables 4.18 and 4.20.

Next to this, significant differences are observed in the structural performance of the two main axes that define configurations 1 and 2, as confirmed by the structural testing and analysis. Following graph 4.16, both the bending strength and stiffness of configuration 2 is significantly lower, comparing to configuration 1. Moreover, in configuration 1 the tube is the element that fails, whereas in configuration 2 the joint presents significant deformation. Based on tables 4.19 and 4.21, in comparison with table 4.8, paragraph 4.2 'Guidelines', the $F_{max,Linear}$ of configuration 1 is higher than this of a single paper-tube, whereas that of configuration 2 is significantly lower.

In a different way, the compression test shows that the bolted joints work effectively as a fixation method between the joint and the tubes. At the same time, it indicates the importance of supporting the paper-tubes from the side, against shear forces, to prevent damage of the paper-tubes in case of excessive axial forces.

- Global observations from 'ANSYS FEM Analysis'

The main conclusion from the 'ANSYS FEM analysis' of the multi-axial joint is that the maximum principle stresses calculated indicate that the joint is sufficiently strong in order to satisfy the boundary conditions set, whereas the condition of the beams is particularly critical. Next to this, further studies about the mechanical fixations are required, both for optimization of the assembly method and also further analysis.

- 'ANSYS' versus Structural testing

Comparison between the 'ANSYS' analyses of the linear configurations and the testing results shows good relevance regarding the maximum stresses calculated at the area of the support (point of force-application F). This identification is positive for the 'ANSYS Numerical analysis' carried out.

On the other hand, as stated before, the idealized conditions within the FEM analysis do not encounter the same stability issues for configuration 2 that rose in the structural experiments.

Graph 4.16 Collective data for case study 'Timber block puzzle node'. Results from 4-point bending tests on configurations 1 and 2 and also paper tubes. To explain the legend, the initials 'MW' were used as a code name for this case study, for the testing process, the number 1 or 2 indicates the configuration.

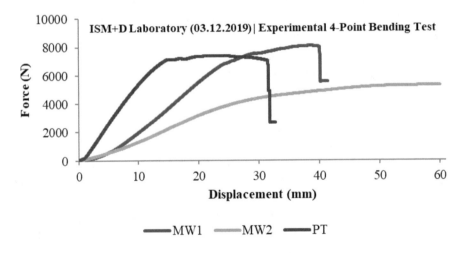

Table 4.22 Overview of results from 4-point bending test (mean values).

Timber block puzzle node	Mean F_{max} Ultimate (N)	Displacement Point A (mm)	Mean F_{max} Linear (N)	Displacement Point A (mm)
Configuration 1	8071	38.45	6500	23
Configuration 2	5305.5	59.76	3500	22

Literature

David W. Green, J. E. W., David E. Kretschmann. (2010). Mechanical Properties of Wood. Retrieved from https://www.fs.usda.gov/treesearch/pubs/37427

KVH. *High-performance building materials for timber construction*. Retrieved from www.kvh.eu

McQuaid, M. (2003). *Shigeru Ban*. London N1 9PA: Phaidon Press Limited, Regent's Wharf All Saint's Street.

Miyake, R. (2009). *Shigeru Ban Paper in Architecture*. US: Rizzoli International Publications.

4.2.1.3 Tolerance adaptive timber plug

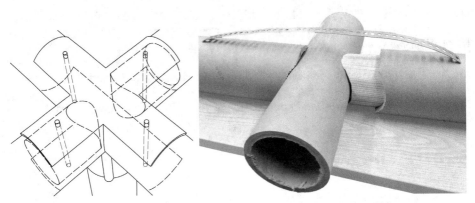

Figure 4.53 (Left) the concept, (right) draft prototype - simplified version of the design.

One of the main aspects addressed during the research on intermediate wooden joints is the complexity caused by the significance of precision in the manufacturing of the structural elements as well as the joints, but also the tolerances within an assembly.

This case study aims to introduce an alternative to the most common plug-in joints examined before. The idea is to simplify the assembly and production methods. Therefore, only a draft prototype was created, as a specimen for a bending test, in order to look a bit closer into the structural behavior of the joining system.

Below, the intended assembly concept is described extensively, following the details sketched in table 4.23. The main characteristic of it is that it leaves one of the beams uninterrupted. This beam is used as a boundary on which all adjacent paper-tubes are form-fitted. This way form-locking mechanisms are an important part in this concept, but at the same time are kept very simple. Hence, they are easy to realize by performing simple cuts, without significant risk for miss-fitting due to imprecisions. An inner wooden plug (fig. 4.55) is integrated in each of the beams (except for the continuous one). This plug is curved on one end so that it shares a contact surface with the continuous beam. It is used to integrate a steel rod that goes through the continuous tube and secures the aligned beams in place (fig. 4.54). The rod is inserted about 40mm in each plug. Then, pressure plates are placed as shown in fig. 4.53 (left) that are bolted together (top to bottom). At the bottom side of the assembly the steel plates shape corners, to ensure stability of the system. Next to this, ideally, these corner plates can be slid into each other (snap fit) to create a firm ring around the supporting column.

Due to the limitation of time it was not possible to examine this case study extensively. The draft prototype shown in fig. 4.53 (right) is a cross-assembly, dimensioned so that it fits the requirements for the experimental 4-point bending test. Instead of steel pressure plates, a perforated steel strap is applied that works as a tensioning element rather than applying pressure. The plugs are bolted together with the paper-tubes and the steel straps.

Table 4.23 Characteristic joining method – 'Tolerance adaptive timber plug'.

	Form-locking		Between tubes and plug
Force-closure	**Friction**		
	Pressure		Bolted Steel plates
	Tension		Bolts + sliding plates
	Inertial Forces		Tensile cables
	Material-closure	-	-

Detailing

An option would be to adjust the form-locking between the tubes, by creating further fittings, such as these shown in concept 2, table 4.1, 4.1.1. On the other hand, an alternative futuristic design, in the spirit of a tolerance adaptive joint, could avoid completely shaping the edges of the tubes and solve the problem in a different way. An idea is to use a compressible material, to cap the tubes and hold them together (design principle 7, table 4.1, 4.1.1). This would need to be laminated on all contact surfaces.

Assembly techniques

Due to the absence of intermediate nodes, either the whole structure shall be prefabricated or assembled on the site. On-site assembly is recommended to avoid damages during transportation. When no glue is applied between the structural elements, then disassembly and relocation of the structure should be possible.

Production and prototyping

The main steps required for the manufacturing are the following: Performing the curved cuts on the tubes (diamond core drill), milling the wooden plugs (combination of basic power tools or use of CNC Milling technology), forming the pressure plates and drilling the holes for the steel threads. From all previous steps, handling the steel profiles is the most difficult operation that requires special skills.

In order to build nodes that use the sliding mechanism mentioned in the introduction, professional manufacturing of the steel elements would be needed. Otherwise, steel plates that are bent in the right curve and welded together would be sufficient.

In the prototyping process performed, the wooden plugs were formed as following: a squared timber profile was used. First the curved surface for contact with the intermediate beam was created. Free-hand routing was performed on both sides of the profile and then a diamond core drill was used to make a clear cut. The pieces were finally sanded. To transform the squared profile to a cylinder the same practices as in the case study 'timber block node' were applied.

 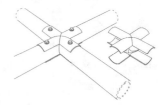

Figure 4.54 (Left) fixation elements, (middle) sliding mechanism, (bottom) version for curved structure.

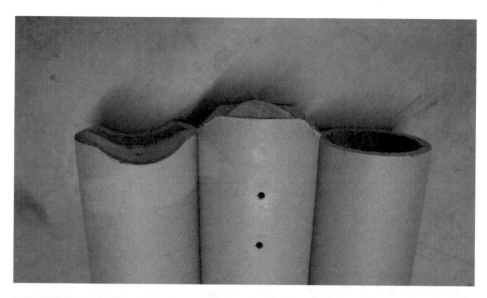

Figure 4.55 Prototyping, (top) assembly-elements, (bottom) timber-plugs.

Structural Testing

Figure 4.56 The testing process.

The connection between the elements is dry. As the middle tube in this cross is empty, the most critical area of the assembly is identified there, around the holes that allow access to the steel thread (4.57, right) and the contact surface between the tubes (fig. 4.57, left). Even though the specimen did not collapse, it showed limited stiffness, as it developed high rotation around the center of the middle tube.

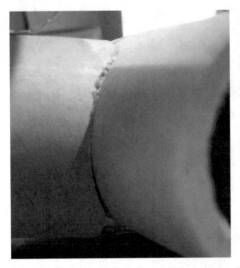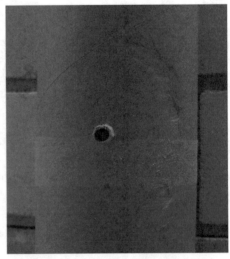

Figure 4.57 (Left) rotation of assembly during testing, (right) impact of testing on the middle profile.

Graph 4.17 'Tolerance adaptive timber plug', testing results, force versus displacement curve. The code name 'TAJ' was used for the testing process.

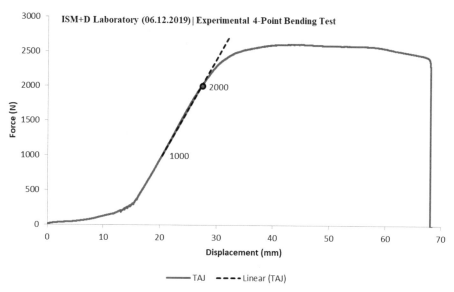

Table 4.24 'Tolerance adaptive timber plug' | F_{max} Ultimate (point A: support, point B: middle).

Specimen 1	F_{max} (N) (Ultimate)	Displacement (mm) Point A	Displacement (mm) Point B	Time (sec)
	2605	44.40	0	534

Table 4.25 'Tolerance adaptive timber plug' | Fmax Linear (point A: support, point B: middle).

Specimen 1	F_{max} (N) Linear-part	Displacement (mm) Point A	Displacement (mm) Point B	Time (sec)
	2000	27	0	329

Summary

The experimental process reported hereby helps to address the potential of this joining technique as a concept. The structural testing showed limited performance due to the absence of pressure plates. Further studies are required to produce improved assemblies and also specimens for structural testing. The recommendations provided regarding the manufacturing methods could be useful in such an attempt.

4.2.2 Sleeve joints

In this section two different cases of joining techniques are investigated. The first one refers to a mechanically fixed sleeve connector and the second one to a laminated joint that is formed around the assembly of tubes, with layers of textile incorporated in a composite.

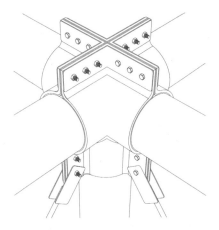

4.2.2.1 Fiber-reinforced pulp-composite connector

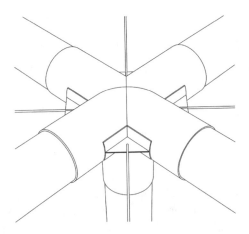

4.2.2.2 Textile-reinforced epoxy resin laminated joint

4.2.2.1 Fiber reinforced pulp-composite connector

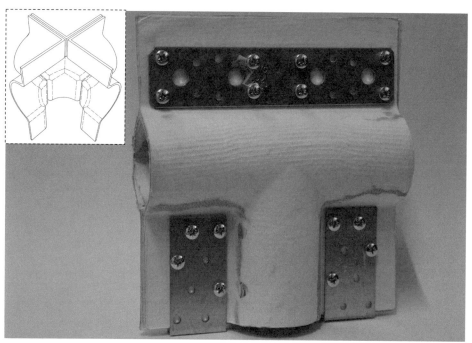

Figure 4.58 T-joint specimen used for the experimental bending test.

This case study has been elaborated in co-operation with researcher Paul Töws (Fraunhofer Institute, LBF), who develops research on fiber-reinforced bio-based composites produced with high-pressure forming (molding). The main concept is to create a sleeve joint that is more eco-friendly comparing to the common solutions and therefore all materials used in the manufacturing process come from renewable sources. These are mainly pulp for the main core of the shell, natural textiles (jute or flax) and starch binder. Common examples of such joining techniques in construction and design projects are made of steel or plastic. The global vision would be to develop alternative multi-axial joints as described above that could be used to generate various assemblies and entire frameworks. Hereby, to make a start, the design of a T-joint is elaborated. The main reason for this selection is the simplicity for manufacturing comparing to other designs. Moreover, this design could potentially be used in various occasions (page 264). The prototypes manufactured for this study are in scale 1:2, following limitations in the mold-making process, as explained in page 265. This fact creates a significant differentiation comparing to the other case studies, limiting thus the comparability, on the level of structural performance, as explained in paragraph 4.3 'Evaluation of case studies'.

Table 4.26 Characteristic joining method – 'Fiber reinforced pulp-composite connector'.

	Form-locking		Between the tubes Clamp fits as a 2nd skin
Force-closure	**Friction**		0 tolerances between connector & tubes
	Pressure		Bolted flanges
	Tension	-	-
	Inertial Forces		Tensile cables
Material-closure			Adhesion between the tubes

Detailing alternatives

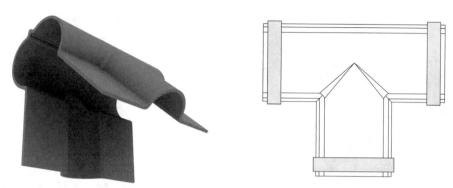

Figure 4.59 Two alternative details for a T-joint, (left) With flexible top edge that works as a hinge, (right) divided in two sides glued along the perimeter and secured with pressure bands.

In table 4.26, the proposed joining method involves the integration of flanges along the edges of the joint, which are reinforced with pressure plates and are bolted together. Another option would be to also laminate the flanges. In fig. 4.59 further alternative details are suggested. The detail on the left could be realized by reinforcing the area of the hinge with fibers oriented accordingly. Next to this, the area of the hinge would be treated slightly different from the rest within the production process, so that it maintains its flexibility. In the detail shown on the right, the bands could either be made of the same material as the joint, textile or metal. To fix them on the joint, various solutions are possible, such as gluing or applying snap or riveted connectors. The detail in fig. 4.59, left, presents advantages for the assembly, whereas the detail on the right offers a very minimalistic design. In both details the absence of a flange along the top edge benefits the potential placement of cladding on top. Contrarily, as a first step for the research process, the bolted flanges offer a steady joint that is simpler to produce successfully, using the design of a composite with thickness about 3mm. Next to this, it is a method that can be easily transformed to nodes with more axes.

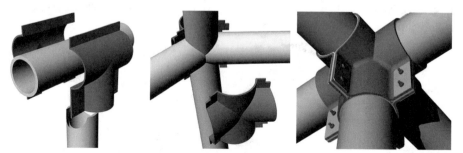

Figure 4.60 A variety of joint-designs is possible, (left) T-joint, (middle) flat or bent cross, (right) multi-axial connector.

Assembly techniques

Figure 4.61 Potential implementation of T-joint in different configurations.

The T-joint allows for many different assemblies, a few of these are shown in fig. 4.61. The main limitation implied by the T-joint has to do with the number of elements that can be connected at one time. This may lead to a high number of joints. Next to this, as a result, an important aspect while configuring any assembly is to prevent eccentricity. To this respect, the arch-tunnel assembly in fig. 4.61 presents some advantages. Additionally, in the case of a rectangular frame, the same T-joint can be used to integrate stiffening diagonals between horizontal beams (top to bottom), right next to the columns. Moreover, the flanges of the joints can be used to attach tensile cables, as shown in table 4.26, as well as a lightweight protective skin.

Considering the small size of the constructions set as a context for this research (chapter 3), this joint presents many advantages for on-site assembly. The main reasons for this are the following. At first, there is a very limited number of different kinds of elements to be assembled. Secondly, the fitting between the elements is more certainly to be successful. The manufacturing process ensures high-precision (clamp) and also the form-fitting between the tubes is easy to manage. The alternative details presented in fig. 4.59 could also be performed in the same spirit. Prefabrication is still a possibility. As it is already addressed, the development of further designs could help to create more minimalistic assemblies. For example, cross-joints (fig. 4.62, left) could be used to reinforce the main grid with secondary beams and also as points to fixate cladding components. Multi-axial joints like this in fig. 4.62 (right). would again simplify the assembly process and expand the potential of this joining method.

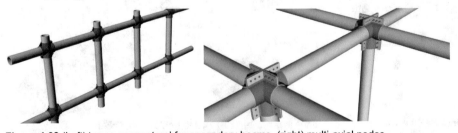

Figure 4.62 (Left) 'cross-connectors' for secondary beams, (right) multi-axial nodes.

Production method and prototyping

In this section insightful frames from the process of manufacturing the shells are described. The creation of the mold was the main obstacle to overcome, in order to manufacture parts. Two different materials were tried UHPC and carbon fiber. The UHPC is the one that could preserve through the pressure-forming.

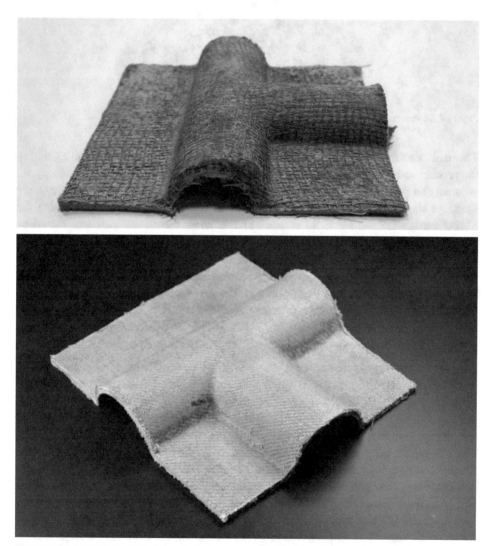

Figure 4.63 Samples from the manufacturing process, (top) 1st series of prototypes – jute fiber, (bottom) optimized series of prototypes – flax fiber (source: Töws, P.).

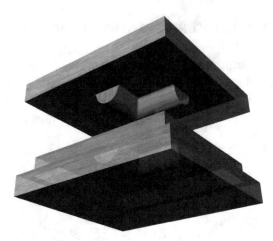

Figure 4.64 Pressure forming – designing the mold.

This mold was first designed in 'Rhino', to control better conditions for handling during the process and also restrict the weight to a maximum 15kg per component (density vs. volume). To form the mold, a few 3d printed copies of the desired final geometry were used. As shown in fig. 4.65, this element was fitted in a wooden mold, in which the UHPC was casted. As the mold consists of two parts, the second part was casted by using the first part and the 3d-printed element on the one side and again a wooden mold. This way, optimal fitting between the two parts is achieved. After the first attempts to form parts, the top part of the mold presented cracks. Thus, a stronger part was produced, this time with open edges around the perimeter to allow the water to drain easier (there is a high percentage of water in it in the start of the process). Moreover, steel threads were integrated in the form (fig. 4.65), as reinforcement. This top part supported the manufacturing of elements throughout the research process.

For the manufacturing process, the first step is to lay the desired stacking in the mold. In this case, the composite is about 3 mm thick. It consists of flax textile on the sides, chemical pulp on top and a layer of mechanical pulp in the middle. In principle a thickness between 1-10 mm is considered possible. For instance, the specimens used to test the performance of bolts are 7 mm thick and then the thickness of the different layers is adjusted accordingly. The orientation of the fibers from the textile is made so that the composite behaves as a quasi-isotropic material. The substance used as a binder is starch-based (wallpaper-glue). Regarding the application of pressure, a force of 30 kN is used to form the simple plates, whereas to form the individual shells for the clamps a force about 10 kN was applied, to prevent damage of the mold. The pressure is applied for about 24 hours (it is an overnight process), so that the composite has enough time to dry and cure. Several production series were performed in order to reach satisfactory results, in terms of rigidity of the shell.

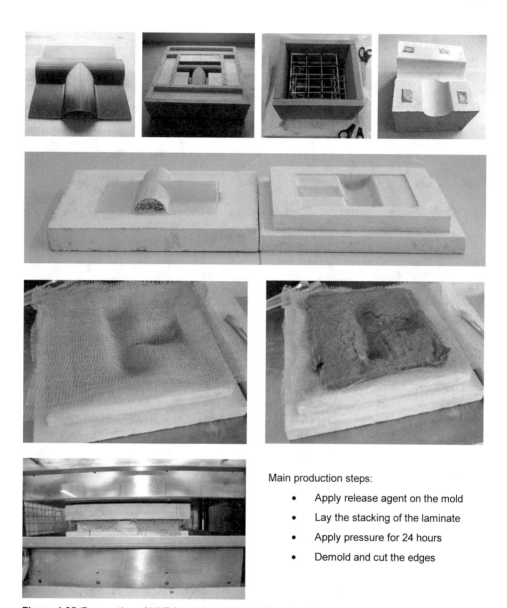

Main production steps:

- Apply release agent on the mold
- Lay the stacking of the laminate
- Apply pressure for 24 hours
- Demold and cut the edges

Figure 4.65 'Preparation of UHPC mold' and 'Production steps'.

The optimized joints were tested for bending, as shown in the next paragraph. Woven flax textile was used for the production. No significant deviations occurred between the volumes of the produced parts or changes due to room-climate, although the parts are not waterproof.

Structural testing

Figure 4.66 Testing set-up. The testing principle is the same as in fig. 6.4 paragraph 6.2. The tubes are made of steel and are only connected in terms of dry form-fitting.

Experimental vertical bending test

This series of testing was performed in the pre-final stage of structural testing. The prototypes are in scale 1:2, as a mold made of UHPC in scale 1:1 would be difficult to handle. Thus, the outer diameter of tubes that fit in the joint is 60 mm. At the same time, no thin-walled paper-tubes that would be strong enough to perform a structural test are available. - The paper-tube would bend and the joint would remain intact, as preliminary tests showed. - The main focus is to examine the behavior of the joint itself. The pipes used to build the set-up are made of steel. Between these only dry form-fitting applies. Hence, as the horizontal steel pipe is free to develop rotational movement, the T-joint is the only element that secures the assembly together. The vertical steel pipe is welded on a steel base (at the bottom of the assembly). The distance between the center of the joint and the location on which the Force is applied is 1 m. The same kind of set-up can be observed more extensively in further series of experiments (see paragraph 6.2, appx.). In total three specimens were tested. All of them failed in the same way. The shell presented high distortion on the front end, where the load causes rotation. This distortion affected mainly the area between the edge right under the pipe and until the start of the steel pressure plate. There, splitting between the stack-up is observed (fig. 4.67). Further progression of the rotation lead to damage of the shells caused by the metal plates. Therefore, the connection between the shells needs to be optimized, for example, by using metal plates that cover the entire surface of the flange. Moreover, laminating the shells together would also be a good reinforcement.

Figure 4.67 Distortion of composite shell due to loading.

Graphs 4.18 and 4.19 Vertical displacement (top) and rotational movement (bottom) versus force.

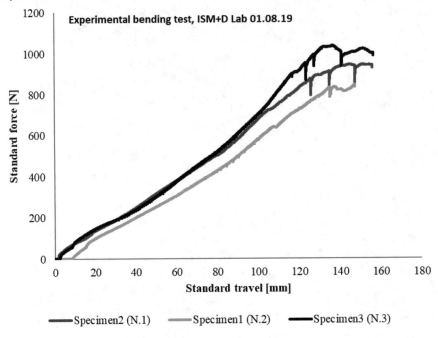

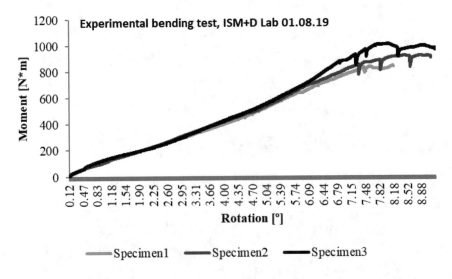

This testing process represents a preliminary experiment and not the structural capacity of a finished product.

Double lap shear test

With this shear test, the effect of shear on the bolts is observed. This is particularly important, for instance, for loads that occur along the same or a parallel plane to the flanges of the T-joint and can cause the two parts of the shell to move in opposite directions. The double lap joint is selected as a testing set-up because of its symmetry, even though it is not identical with the system expressed in the T-joint.

Hereby the main details for the specimens that are tested are addressed. The individual plates are 7 mm thick. The material is composed of 2 mm flax textile on the sides, 2 mm chemical pulp, 3 mm mechanical pulp in the middle and another 2 mm chemical pulp. The size of the bolts is M6. The dimensions of the individual plates are 50 mm*150 mm. The distance of the bolt from the sides of the plates is 20 mm, based on guidelines for fiber-reinforced composites (Schürmann, 2005). The Force is applied with a speed of 10 mm/min.

A combination of failing mechanisms can be observed. It is mainly a result of shear around the bearing (bolt) that causes the hole to deform and tension parallel to the edge (see disassembled specimens in fig. 4.68) that strains the fibers of the textile. The difference between the curve of specimen F comparing to the rest is related to the torque applied on the bolts only. Evaluating the results that regard the majority of the specimens and express the same pattern (specimens A, B, C and E) a load of roughly 100 kg seems to the maximum force that can be carried per bolt before plastic deformation starts.

Further testing is required to identify the optimal thickness and matrix of the composite for the T-joint, before reaching the point of selecting values that represent the performance of the bolted joint. Moreover, the material properties of the composite are being extensively evaluated by its developer Paul Töws. For instance, for the woven flax textile a tensile strength of 100 MPa and Young's Modulus 6 GPa are reported. Additionally, with use of unidirectional flax textile the same values rise to 210 MPa and 13 GPa respectively.

Graph 4.20 Shear test, double lap bolted joint (fiber reinforced pulp-composites plates).

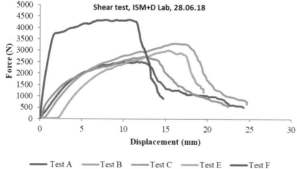

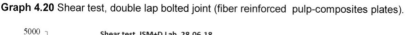

Figure 4.68 Set-up for shear tests and failed specimens.

Summary

This study succeeded to get a first glimpse on the process of designing, manufacturing and understanding the structural behaviour of the examined bio-based sleeve joints. Still, the research process reported indicates that many aspects should be further developed to reach the desired performances and generate a result that reflects the potential of the composite material in use.

The design of the shells functions properly, as does the fitting of the elements produced.

One of the main issues encountered is that the connection between the two shells needs to be re-developed and tested, as described previously. This can still be done with the 1:2 scale prototypes.

For proper structural testing, full-scale joints with higher wall thickness are necessary. These could be tested similarly to the cross-specimens presented in the next case study about fiber-reinforced laminated joints. Moreover, the FEM analysis developed in that same study could be viewed as a guide for the performance of this joint as well, due to the many similarities between these two cases.

Literature

Schürmann, H. (2005). *Konstruiren mit Faser-Kunstoff-Verbunden*: Springer Berlin Heidelberg New York.

Acknowledgments

I would like to thank researcher Paul Töws for his contribution to this case study by implementing his findings on this special application and for offering his expertise in manufacturing the shell parts. Next to this, I would like to thank Marcel Hörbert, for providing his skills in processing steel products, to build the testing set-up.

4.2.2.2 Textile-reinforced epoxy resin laminated joint

Figure 4.69 Manufacturing the final specimens for the experimental 4-point bending test.

This case study examines the possibility of fiber-reinforced adhesive joints that are directly formed on the paper-tubes, following a simple hand-layup process that does not required any sophisticated equipment. There are a few inspiring references, for instance bike frames constructed with bamboo that present this technique as a craftsmanship with great potential for building frames with high aesthetics. The first experiments were carried out together with the guest-student Noah Gilsdorf (appx. 6.4.1.3), within the context of an international research exchange program (IREP). Then, as a starting point, the focus was on forming small-scale prototypes for a variety of assemblies. The first important aspect was to examine the compatibility of this joining method with paper-tubes. As the findings from the prototyping process were positive, the next goal was to gain experience in designing the lay-up, identify what is the best way to use the textile and yarn within the layering process and plan assemblies that are more convenient to manufacture in a short time, following the time-span for which the average resin product is workable. As the results from this process were encouraging, after the completion of this program, the experiments presented hereby were elaborated, to examine this possibility in a context that is comparable with the other case studies.

Table 4.27 Characteristic joining method – 'Textile-reinforced epoxy resin laminated joint'.

	Form-locking		Between the main struc-tural elements
Force-closure	**Friction**	-	-
	Pressure		From the shell on the structural system
	Tension		Fiber/ textile nets
	Inertial Forces		Laminated steel profiles
Material-closure			Between tubes + Cured composite

Detailing

This joining method allows for a broad variety of assemblies (see paragraph 6.2, appx.). Simple contact (form-fitting) between the structural elements is required as a base for a successful assembly. The first factors that can be adjusted, depending on the design requirements, are the wall-thickness of the shell and length of 'sleeve'. For example, higher length would be required for beams that cover longer spans, to reinforce the assembly. These features have great impact on the production process, meaning both the amount of material and time required. One of the main priorities while designing the individual layers for the shell is to integrate continuous layers of textile as a bridge between the individual members and to reinforce the areas that are critical for bending. Local reinforcement of the later ones is possible and can be very efficient, as by minimizing the surface of textile used, the amount of adhesive applied is also eliminated. The matrix of the shell defines the structural performance. The main difficulty while designing a lay-up that is non-uniform is the difficulty in predicting the structural performance, either with simulations or with structural testing. The development of the matrix itself is used to reinforce the shell. Local reinforcement can be enhanced with extra layers of adhesive that may be applied subsequently, after the joint has cured. Furthermore, the integration of metal sheets in the assembly is also an option, with the downside that disassembly is not possible. A different approach would be to integrate diagonal beams in close distance to the joints and connect them in the same way.

Assembly techniques

As mentioned beforehand this joining method is easier to be hand-crafted. Thereby, the erection of the structure needs to be phased. From the start, temporary supports are required, to set-up the frame or parts of it. For every part of a frame that is produced, first the paper tubes need to be attached, with a glue bond. Then, the handcrafted joints can be created, as explained in the next paragraph about production. Then the joints can be formed and the temporary supports can be removed only when these have fully cured. The structure needs to be produced in close proximity to the area of installation and should not be relocated. Acknowledging the difficulties that may rise in the previous process, some alternatives for pre-manufacturing construction elements are considered. An option would be to fabricate shells, similarly to the case study 4.2.2.1 and then laminate them on the tubes. If the frame would be entirely prefabricated then perhaps it's even possible to integrate the beams in the molding process. This assembly method is only conceptual, as many details within the production method would need to be worked out for realistic implementation. Next to this, it would lead to a costlier process. An important aspect to address is that disassembly of the frame is not possible and recycling of the joints could only be possible when water-soluble adhesives would be used. This is a subject that requires further investigation.

Production

Figure 4.70 A supporting structure is always necessary for easy handling during forming.

In this section the most important aspects that need to be worked out in order to perform the forming process are explained.

It is essential to create some supports for the tubes that elevate them from the main working surface, as shown in fig. 4.70. For the current experiments the supports were simply formed out of timber plates that allow to rotate the assembly comfortably, to work on both sides of the prototype. More sophisticated supportive frames could be made with aluminum profiles. This idea would work great for more complex structural systems with bigger size, as it would provide stable grip.

A particularly crucial aspect for the success of the process is the proper use of the adhesive. In this case epoxy resin is used. This is composed of two elements, the resin and the hardener. The one critical point is to mix these parts with the exact proportion as prescribed, using always a clean bucket and mixer, to ensure that the mixture will function properly. The other critical point is to realize the assembly process within the timeframe for which the workability of the resin is high (30' in this case). In principle, it is possible to select resin mixtures that allow for extension of this timeframe to a couple of hours and then accelerate the curing process when needed. An essential part of the process is that after the matrix of the composite is formed, the air between the layers needs to be removed. For this purpose, tape is used. The right kind of tape shall be chosen. It should be possible to remove after a day. Insulating (or electrical) tape works for this purpose.

Figure 4.71 The main layers required for the assembly of the cross-joint (configuration 2).

As part of the experiments, to produce the specimens needed for structural testing, two kinds of prototypes were worked-out: simple linear laminates between two tubes and a cross-assembly. The first aims to define the strength of the shell itself and the second one the strength of a joint that is based on form-locking between three elements. This assembly creates more critical conditions as it involves two seams, it increases the possibility of contact issues between the different parts due to the non-uniform thickness of the tubes along the contact lines.

Following the experiments, the main steps required for a successful lay-up process are the following. The textiles need to be pre-wetted, using a brush. Then all layers are placed on the tubes one by one, each time applying first resin on the bottom layer.

For the prototype shown in fig. 4.70, four layers of textile overlap at maximum within the thickness of the shell. These are shown in fig. 4.71. It is important that these layers share common surface, so that they bind strongly. The thin stripes are applied in the end, to reinforce the inner corners that are formed around the cross.

During this process, ropes may be used to temporarily tighten the layers of textile together. Before applying the pressure-tape these shall be removed, to prevent air-bubbles in the shell. Moreover, it is important to apply equal pressure while tensioning the tape around the prototype.

In principle, the more resin that is used, the stronger the shell becomes. However, keeping in mind that resin is a costly and also unsustainable product, it is best to use the least amount possible.

When the resin has cured and the tape is removed the final surface of the composite will be rough and might present wrinkles. To improve the smoothness of the surface it is possible to sand these down. In this case it is very important to not inhale the dust. Then the joint may be coated with an extra layer of resin. Coloring of the resin for special aesthetical effects is also possible.

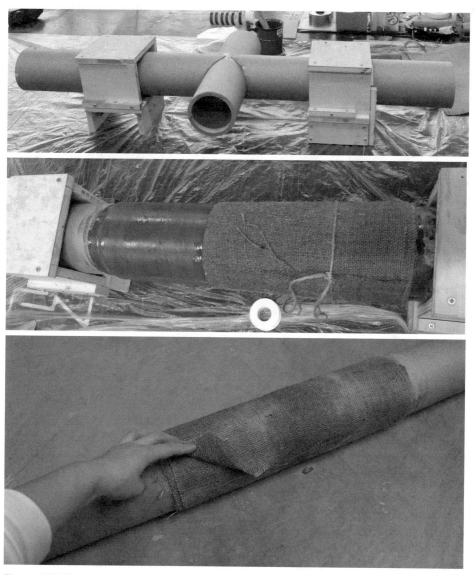

Figure 4.72 (Top) forming process, step 1, (middle) use of special tape to apply compression, (bottom) failed forming process due to issues with the resin mix prevents the curing process.

Structural Testing

Testing configuration 1

Figure 4.73 Configuration 1 (specimen 2), the testing set-up.

The specimen shown in fig. 4.73 is composed of two tubes, glued on their contact surface and connected with four continuous layers of jute textile hardened with epoxy resin.

The resin product used is 'Presto'. The amount of resin used per specimen is about 250grams.

Then mean $F_{max,Linear}$ is about 5 kN (table 4.29), with a respective vertical displacement of 7 mm approximately. The Mean $F_{max,Ultimate}$ is about 6,88 kN (table 4.29), with a respective vertical displacement of 19.5 mm approximately. The margin between the minimum and maximum $F_{max,Ultimate}$ is about 535 N. The maximum displacement measured by the sensor for the same spectrum of values is 0.73 mm.

All three specimens presented the same failure mechanism. The specimen raptured on the bottom side of the paper-tube, right next to the finishing edge of the joint. More specifically, the delamination occurred between the layers of the paper-tube, as expected. The composite shell didn't present any signs of delamination. However, specimen 1 (fig. 4.74, top) presented rupture also on the joint, that was initiated from the outer edge towards the inner core of the composite shell. As the textile splits easily on the outer edges it is important to make sure that these are coated as well.

Following this series of testing, it is a characteristic of this joining method not to provide significant warnings for the upcoming failure.

Graph 4.21 'Textile-reinforced epoxy resin laminated joint', testing results, force versus displacement curves. To explain the legend, the initials 'LAM-T' were used as a code name for this case study, for the testing process, 'c1' indicates the number of the configuration and the numbers 1,2 and 3 the different specimens.

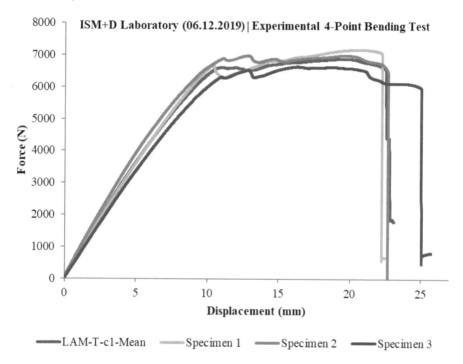

Table 4.28 'Textile-reinforced epoxy resin laminated joint' | Testing configuration 1 | F_{max} Ultimate (point A: support, point B: middle).

Configuration 1 Specimen Nr.	F_{max} (N) (Ultimate)	Displacement (mm) Point A	Displacement (mm) Point B	Time (sec)
Specimen 1	7168	20.80	0.73	250
Specimen 2	6995	19.70	0.69	237
Specimen 3	6633	16.28	0.48	196

Graph 4.22 'Textile-reinforced epoxy resin laminated joint', force versus displacement average curve. To explain the legend, the initials 'LAM-T' were used as a code name for this case study, for the testing process and 'c1' indicates the number of the configuration.

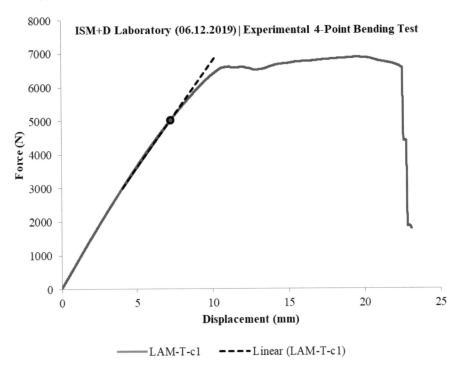

ISM+D Laboratory (06.12.2019) | Experimental 4-Point Bending Test

——LAM-T-c1   ----Linear (LAM-T-c1)

Table 4.29 'Textile-reinforced epoxy resin laminated joint', testing configuration 1, selected mean values.

Configuration 1, Mean values	Mean F_{max} Ultimate (N)	Displacement Point A (mm)	Mean F_{max} Linear (N)	Displacement Point A (mm)
	6883.7	19.45	5000	7

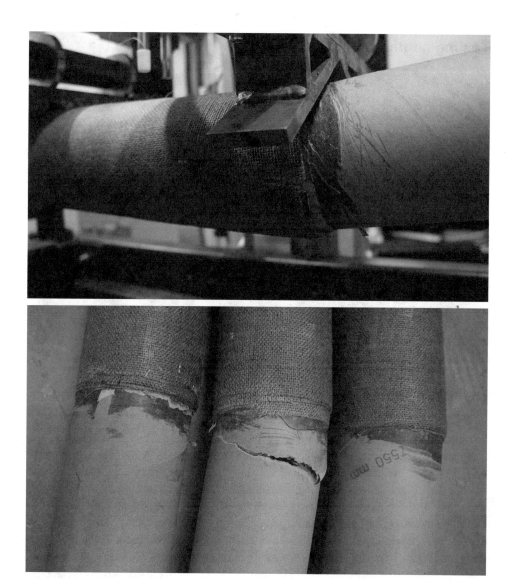

Figure 4.74 (Top) testing process – specimen 1, (bottom) failed specimens - specimens 1 to 3 from left to right.

Testing configuration 2

Figure 4.75 Testing configuration 2, failed specimens.

In total three specimens were tested. The resin product used for specimens 1 and 2 is 'Presto' and for specimen 3 'AutoK'. The amount of resin used per specimen is about 400grams.

The mean $F_{max,Linear}$ is about 4,3 kN (table 4.31), with a respective vertical displacement of 7,8 mm approximately. The Mean $F_{max,Ultimate}$ is 6 kN approximately, with a respective vertical displacement of 15,5 mm approximately. The maximum displacement measured by the sensor for the same spectrum of values is 1,56 mm. The margin between the minimum and maximum $F_{max,Ultimate}$ is 1,5 kN approximately. This high difference is, to a certain extent, related to the different resin products used. To be more specific, specimens 1 and 2 present performances that are comparable, as it can be seen in table 4.30, whereas specimen 3 is significantly stronger. As there are various parameters that affect the performance it is not certain if this higher strength is a result of slightly better overlapping between the layers of textile, more successful removal of the air or the resin mixture. Therefore, further experiments are required. This time, the specimens fail on the side of the joint. The composite shell presents cracks. This result is mainly a combination of two factors. The one is that the layers of textile are not continuous like in the previous specimens but split in patches that are bound in a single body. The other factor is the limited strength of the glued joint between the tubes. For this joint wood glue (Ponal is used). As it can be seen in fig. 4.76 (bottom). the composite shell fails along one of its weak paths, on the bottom, close to the center. As the textile has absorbed the resin, the shell remains in one body, with no signals of delamination. The crack propagation doesn't seem to follow always the same pattern but could be related to inhomogeneities and weaker areas in the composite. Again, no significant warnings for the upcoming failure were noticed.

Graph 4.23 'Textile-reinforced epoxy resin laminated joint', testing results, force versus displacement curves. To explain the legend, the initials 'LAM-T' were used as a code name for this case study, for the testing process, 'c2' indicates the number of the configuration and the numbers 1,2 and 3 the different specimens.

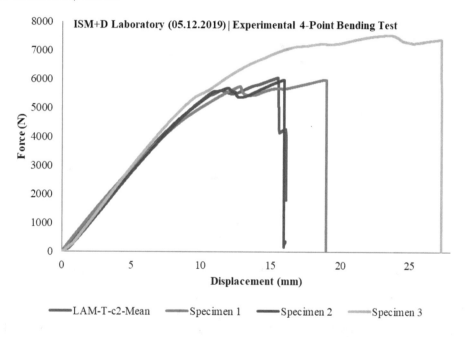

Table 4.30 'Textile-reinforced epoxy resin laminated joint' | Testing configuration 2 | Fmax Ultimate (point A: support, point B: middle).

Configuration 2 Specimen Nr.	F_{max} (N) (Ultimate)	Displacement (mm) Point A	Displacement (mm) Point B	Time (sec)
Specimen 1	5985	18.90	1.23	228
Specimen 2	5977.5	15.92	-0.38	192
Specimen 3	7535	23.44	1.56	282

Graph 4.24 'Textile-reinforced epoxy resin laminated joint', force versus displacement average curve. To explain the legend, the initials 'LAM-T' were used as a code name for this case study, for the testing process and 'c2' indicates the number of the configuration.

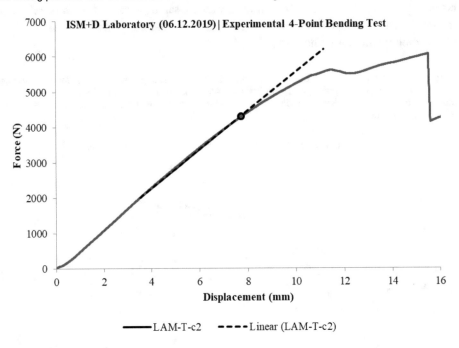

Table 4.31 'Textile-reinforced epoxy resin laminated joint', testing configuration 2, selected mean values.

Configuration 2, mean values	Mean F_{max} Ultimate (N)	Displacement Point A (mm)	Mean F_{max} Linear (N)	Displacement Point A (mm)
	6055,8	15,5	4300	7,76

Figure 4.76 (Top) testing configuration 2, set-up (specimen nr. 1), (bottom) failed specimen - nr. 3.

Local FEM Analysis

Multi-axial node

The geometry and principle assembly of the joint are as presented in table 4.27. The ANSYS model, including the load conditions, is shown in fig. 4.77 (top). An important simplification made is that the composite shell is assumed to have uniform thickness and is designed as one body. The contact property between the shell and the tube is bonded. Another simplification regards the implementation of steel cables that is realized with steel profiles that are laminated on the shell. Hence, the contact property applied is bonded.

The material properties assumed, for the composite shell, follow. The applied properties are derived from some studies on vacuum formed jute-epoxy composites (Ferreira, 2016). The values estimated for flexural modulus and flexural strength (graphs presented in tables 6 and 5 of the referenced source) are hereby used to define the young's modulus of the composite material, within the model, and to evaluate the results on maximum principle stress respectively. This way, a flexural modulus of 3.5 GPa is assumed, based on the graph presented in table 6 of the referenced publication. This value is meant for a medium woven mat (grammage), composed of four layers of textile and after it's exposed to water. Even though exposure to water is not encountered in the current case studies, to be on the safe side, the value on the lower end is selected. In the same spirit the flexural strength is defined at 70 MPa.

The mechanical behaviour could be significantly better when other kinds of fibers, such as flax, would be used instead (Bougherara, 2014).

The Poisson's ratio is 0.3, according to the properties of the resin (Michigan), (Epoxy-technology).

A global impression regarding the distribution of stresses is provided in fig. 4.77 (bottom). As expected, the peak stresses occur on the plates that fixate the tensile cables on the composite joint. In principle, concentration of stresses occurs:

- at the areas surrounding the connected edges between the paper tubes
- the areas under tension, following the bending of the tubes
- at the boundaries of the joint, where the system becomes suddenly weaker

In fig. 4.78, exports from the evaluation of the distribution of stresses for the different parts of the geometry are provided, and more specifically, the outer and inner surface of the composite shell and the paper tubes. In these images the general observations stated before can be observed clearly. The range of values presented in these evaluations is also interesting to observe. At the same time, it is important to keep in mind that due to the simplifications regarding the mechanical behaviour of the materials, the resulting stresses should be significantly lower than the limits. Within the composite shell, the range of values is between -1.07 and 5.72 MPa (fig. 4.78 top and middle), which in principle is a range of values that lies within the capacity of the material. The range of stresses that occur on the paper tubes is between 0 and 3 MPa approximately (fig. 4.78, bottom), also theoretically within the capacity of the material.

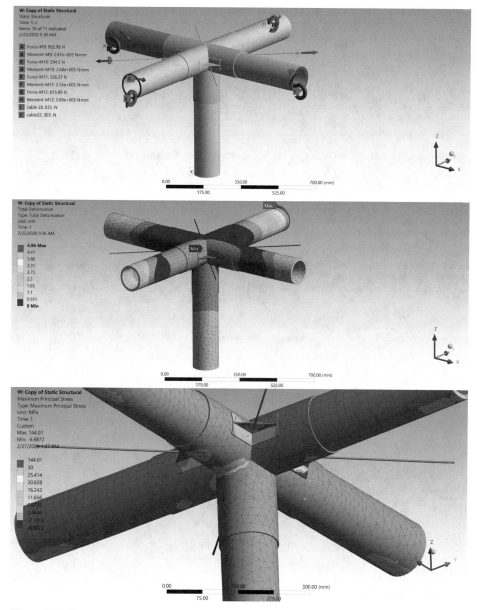

Figure 4.77 (Top) geomtery, (middle) total Deformation, (bottom) 'Maximum Principal Stress'.

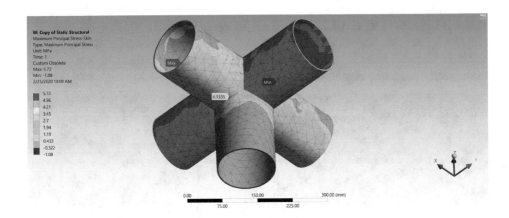

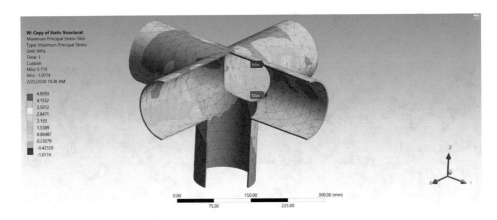

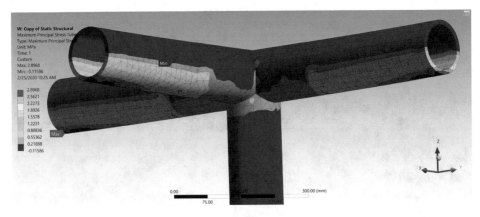

Figure 4.78 Distribution of stresses, (top) outer surface of shell, (middle) inner surface of shell, (bottom) paper-tubes.

Configurations 1 and 2

The 3d models follow exactly the design of the specimens for the respective structural tests. The simplifications mentioned for the analysis of the multi-axial design apply also here. The same goes for all the conditions set for the maximum principal stress analysis. The main difference regards the application of loads. These are equal to the $F_{max, Linear}$, following the results from structural testing and can be seen in figs. 4.79 and 4.80 (top).

It is worth to mention that the comparison between the idealized 3d models examined hereby - where the condition of a uniform thickness applies for the thin-walled composite-, with the real case, as explained in the production process and the structural experiments, is particularly interesting. This way, the full potential of this joining method is pictured, next to the results of simply hand-crafted prototypes. Ideally, optimization of the matrix of the thin-walled composite would approximate the behaviour of a quasi-isotropic material, like in study 4.2.2.1 'Fiber reinforced pulp-composite connector'.

- Configuration 1

The pattern of the distributed stresses expressed in the results of the maximum principal stress analysis, in fig. 4.79 (bottom), caused by the lateral forces, seems to be quite effective. It is a result of the uniform bonded connection between the tubes and the joint.

The peak stresses are demonstrated at the areas where the force is applied. In principle, as expected, higher concentration of stresses occurs at the bottom half of the joint. Next to this, the pattern of pressure occurring on top versus tension at the bottom seems to strain the composite in the middle. The same observation was made also in the process of structural testing, as proved in fig. 4.74 (top), where a crack occurred within the same area. The maximum stress at the area of the support is 6.5 MPa approximately, in close proximity to the value calculated based on the results from the experimental bending test (see 4.3, table 4.37). The maximum total deformation is 9 mm approximately, also in close proximity to the testing results (7 mm).

- Configuration 2

Based on the results of the 'maximum principal stress analysis' the range of stresses is between -9.5 and 11 MPa approximately (fig. 4.80 bottom). As expected, the negative peak stresses occur where the force is applied and the maximum on the opposite side. High concentration of stresses occurs at the bottom of the joint, in the middle area. There, the effect from bending is maximized, when at the same time this is the weakest area of the joint. The thickness of the tubes at the top and bottom is limited and the surface of the elements very narrow. The tubes suffer compression and the composite tension with the center of the junction being the most critical area. It is suspected that the combination of stresses, as a result of tension and compression at these areas, could lead to failure with cracks starting at the lower areas and developing through the corners. Something similar to this scenario happened during the experimental structural test (fig. 4.76, bottom).

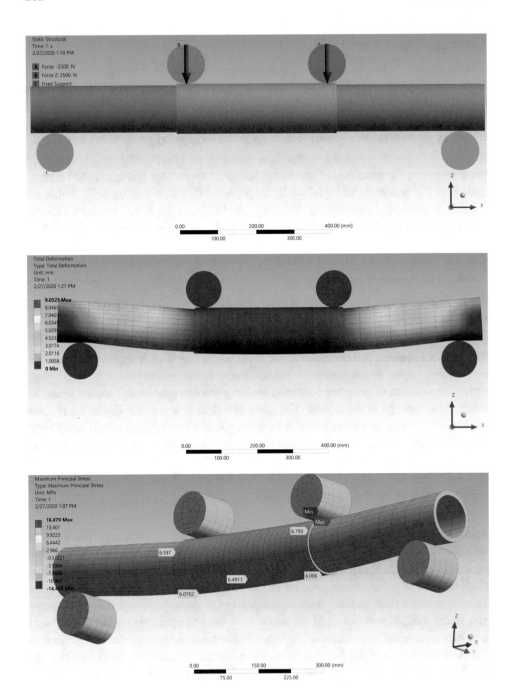

Figure 4.79 Configuration 1, (top) Forces, (middle) Total deformation analysis, (bottom) 'Maximum princpal stress' analysis.

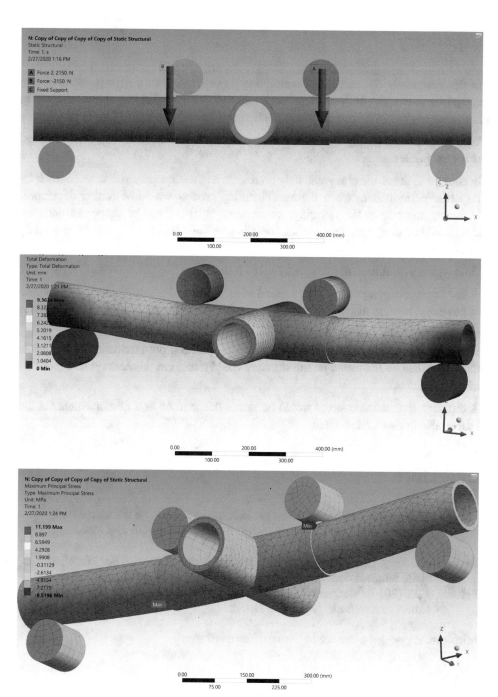

Figure 4.80 Configuration 2. (top) Forces, (middle) Total deformation analysis, (bottom) 'Maximum princpal stress' analysis.

Summary

In this case study the concept of connecting paper-tubes with a fiber-reinforced resin composite is examined. Great steps were taken in the manufacturing process, creating both a variety of assemblies and also transferring a few concepts in full-scale prototypes.

With the structural testing, the results obtained by the linear assembly represent the ideal structural performance.

In contrast to these, the cross-assembly represents a realistic scenario that could reach by far better performances when the manufacturing process would be further developed. Moreover, to optimize the assembly, a continuous layer that can be wrapped around the tubes could be designed. Next to this, the adhesive bond between the paper-tubes could be improved.

Other aspects for further development are the following:

A key to transforming this joining method to an eco-friendly solution would be the use of bio-based binders that would improve both the safety for application (non-hazardous chemistry) and also the 'end of life' phase.

Regarding the optimization of the manufacturing process, pre-fabrication of the joints that could then be laminated on the structure could transform this concept to a more attractive solution.

A different application concept would be to use this method as a reinforcement for the beams, also in case of damage.

Literature

Bougherara, H. (2014). Influence of the Manufacturing Process on the Mechanical Properties of Flax/Epoxy Composites. *Biobased Materials and Bioenergy, 8*, 1-8. doi:10.1166/jbmb.2014.1410

Epoxy-technology. Understanding Mechanical Properties of Epoxies For Modeling, Finite Element Analysis (FEA). Retrieved from www.epotek.com

Ferreira, J. M. (2016). Mechanical Properties of Woven Mat Jute/Epoxy Composites. *Materials Research, 19*(3), 702-710. doi:http://dx.doi.org/10.1590/1980-5373-MR-2015-0422

Michigan. Properties of Seleced Matrices. Retrieved from http://www.mse.mtu.edu/~drjohn/my4150/props.html

Chapter 4.3 Evaluation of case studies

In chapter 4, the experimental research performed, with the aim to examine closely a number of joining techniques for beam structures composed of paper-tubes, is presented. In the start of this process, as shown in table 4.1 'Design concepts for paper-tube beam structures', a spectrum of possibilities is considered, based on the review concluded in paragraph 2.4 'Global overview of joining techniques for paper structures' and the preliminary experiences in this research, as described in paragraph 3.1.1 'an introduction to paper structures'.

For the 'Selection of case studies', paragraph 4.1.2, the decision on the choice of specific research subjects is argued based on the following criteria: transferability of forces, relative cost, assembly process, durability and recyclability. The aim of the tables 4.3, 4.4 and 4.5 'Qualitative evaluation of design concepts' is to outline the global potential of the concepts presented in table 4.1, based on the aforementioned parameters. The concepts selected are these that combine potential for developing joints which are stronger than the tube itself, but with durability comparable to this of paper-tubes, feasible production process (also within the research), relatively simple assembly method that leaves room for minor errors in the forming process (tolerances), with potential for recyclable assemblies. The examples available in the cases of 'timber plate puzzle node' and 'timber block node were considered as an advantage in the selection of these concepts, to further examine the functionality of these joints through this research.

The case studies elaborated (see table 4.32) are divided in the two categories of plug-in (section 4.2.1) and sleeve joints (section 4.2.2) as these are defined in paragraph 4 'Guidelines' (principle joining methods), before the presentation of design concepts. The main aspects elaborated per case study follow, in principle, the criteria discussed in paragraph 4.1.2 for the 'selection of case studies'. These are namely: detailing of the joint, assembly method, prototyping and structural performance (tests and simulations).

The contents of this section, including the observations and evaluation regarding the case studies examined, are organized accordingly, as shown in fig. 4.81. Specifically, the central aim of the evaluation that takes place hereby is to present the key-points and the outcomes from the experiments, to draw comparisons between the different potential solutions and also reflect on the research process and methods used as part of it. For this reason, the pillars of evaluation follow the main aspects analyzed per case study and also consider the criteria set beforehand for the selection of design concepts regarding the durability of the joints. The realization of prototypes has been a fundamental element within the experimental research process for the evaluation of both the design and structural aspects.

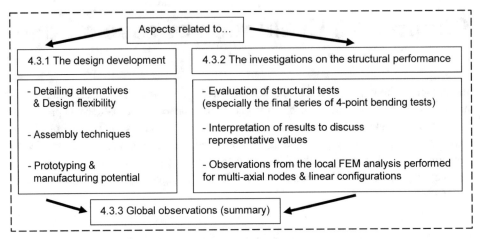

Figure 4.81 Analysis of the outcomes from the elaboration of case studies.

Table 4.32 The case studies of joints elaborated and discussed in chapter 4.2.

Full Name	Timber plate puzzle node (fixed with screws)	Timber block node (fixed with bolts)	Tolerance adaptive timber plug	Fiber-reinforced pulp-composite connector	Textile-reinforced epoxy resin laminated joint
Joining technique (design outlines)					
Typology of joint*	Plug-in			Sleeve	
Case study, paragraph	4.2.1.1	4.2.1.2	4.2.1.3	4.2.2.1	4.2.2.2

4.3.1 Design aspects

Detailing alternatives & design flexibility

All case studies elaborated present great possibilities for the implementation in different assemblies, as achieving design flexibility is one the main goals of the design process. Still, it is not a quality that can be evaluated in a single manner. Considering the experiences gained within the experimental research process, and the insights created regarding the manufacturing process and the structural performance of the case studies examined, the following observations apply:

- For the relation between design and structural integrity

For the category of plug-in joints, the design varies in the three axes (x, y and z). As proved by the structural tests, some differentiations in the structural performance need to be considered when implementing the joints in construction. Recommendations are provided in the respective paragraphs per case study. To summarize the facts, in the case of 'timber plate puzzle node', one of the directions is significantly weaker for bending and therefore is recommended to integrate accordingly in the global structure (for example to absorb compressive forces instead). In a similar way, 'timber block node' presents its highest performance for bending in one direction, parallel to the single timber block, whereas in all other directions, where timber blocks are connected on the primary axis of the joint, the performance for bending is significantly decreased. This problem can be resolved with design optimization and then it needs to be reexamined with further testing. Based on the composition of the nodes, 'timber plate puzzle node' presents structural advantages for beams with smaller diameter, whether 'timber block node' suits better for the design of joints with higher diameter. This observation is based on the fact that the bigger the surface of the node, the higher the bending effect in the timber plates of the puzzle node, whereas the connections between the timber blocks, in the second case, would be more stable and the other way around.

In the case of the 'tolerance adaptive timber plug', the designs shown in fig. 4.54, with the pressure plates clamped around the tubes would be the best version of the design to ensure satisfactory results, by protecting the paper-tubes and providing reinforcement.

In the case of 'fiber-reinforced pulp-composite connector', the simplicity of the design makes it easily applicable for almost any configuration. The optimization of the matrix with higher shell-thickness and implementation of wider pressure plates to prevent deformations are important to improve the structural performance. At the same time, these adjustments require minor changes in the design. However, as discussed later, the manufacturing process is the challenge in this case.

The 'textile-reinforced epoxy resin laminated joint' provides great freedom in the design process, as the textile is used in patches that are laminated together. The strength of the shell can be improved with minimum impact on the design.

- For the relation between design and manufacturing process

The feasibility of the different designs of nodes depends highly on the manufacturing processes, as required for the building process. Thus, the aim is to highlight the impact of the manufacturing process on the range of applications for the different joints. When the forming processes required are easily accessible (such as use of common power tools or automated processes like CNC milling), then it is possible to proceed with adjustments in the design of the components. Instead, when special equipment is required, such as a new mold, as in the case of components produced with high-pressure forming, then decisions for altering the design or proceeding with production of different versions require further evaluation. To this respect, following the suggested manufacturing processes per case, the 'plug-in' joining methods examined in paragraph 4.2.1 are, in principle, easier to implement in different designs, for systems that require different angles or number of axes, comparing to the sleeve joints studied in paragraph 4.2.2. For the 'fiber-reinforced pulp-composite connector', paragraph 4.2.2.1, the manufacturing process is the main obstacle in characterizing the design of the joint flexible, whereas 'textile-reinforced epoxy resin laminated joint', paragraph 4.2.2.2, presents great advantages.

- For the transferability of the joints to beam structures of various configurations and specifically the three typical cases discussed in paragraphs 3.1.3 and 3.2.

For all case studies examined, the design can be altered to fit in the outlines of the global construction and the different nodes occurring in the whole structural system. For instance, case studies 'timber plate puzzle node' and 'textile-reinforced epoxy resin laminated joint' could be implemented in all typical cases.

Joining method 'timber block node' cannot, for the moment, be most effectively implemented in a dome structure (typical case 3), as described in this research, following the differentiations in the performance between configurations 1 and 2. For this reason, the design optimizations suggested beforehand would be necessary.

Case study 'fiber-reinforced pulp-composite connector' would be more successfully implemented in structures where limited lateral forces and moments occur, such as 'typical cases 2 and 3', rather than 'typical case 1'. The same would apply for case study 'textile-reinforced epoxy resin laminated joint', especially for reasons related to the manufacturing process that requires supports to hold the beams in place until the joints have cured.

Focusing purely on the design aspect, the 'tolerance adaptive timber plug' is a solution created for serving different set-ups, but further experimentation is required.

- For the compatibility with reinforcement methods, as discussed in paragraph 3.2 'global FEM analysis'.

Joints with an intermediate core (4.2.1.1 'timber plate puzzle node', 4.2.1.2 'timber block node') or with flanges on the sides (4.2.2.1 'fiber-reinforced pulp-composite connector'), provide suitable conditions for this. On the contrary, in the case of the 'textile-reinforced epoxy resin laminated joint' (4.2.2.2) the fixation for the cables creates high concentration of stresses on the side of the composite, as the local FEM analysis indicates (fig. 4.77, bottom and 4.78, middle).

- For the design aesthetics expressed in the different studies.

The case studies examined in this research are mostly evaluated for their functionality on different levels. Still, the aspect of aesthetics cannot be overlooked. In this context, it is mainly defined by the geometrical characteristics of the different joints, but also the texture of the materials used in each case.

Regarding the geometrical characteristics, plug-in joints are hidden for the most part, in the paper-tubes, leaving the mechanical fixations visible. Still, the component, in the center of the node presents sharp edges. The 'timber block node' creates the impression of a uniform component, which is an advantage, comparing to the multi-layered 'timber plate puzzle node'. In a different way, the sleeve-joints (4.2.2) present a different design approach, as they create organic forms, following the shape of the tubes.

In both categories, this of the plug-in and that of sleeve joints, a variety of materials with different aesthetical qualities are available. For instance, different kinds of wood for the plug-in joints and also different kinds of textures for the composite materials that form the sleeve joints. Still, 'sleeve' joints present advantages, as they form a smooth skin around the joint that makes it is easier to achieve a design with minimalistic appearance. Moreover, the appearance of the joint can be highly influenced, as colouring can be added and also post-processing can provide great finishing.

Assembly techniques

The main parameters encountered for the qualitative evaluation of the case studies, as expressed also in table 4.33, regard the aspects of prefabrication, workability, maintenance, reversibility, reuse and critical areas. Below observations for the aforementioned aspects are provided.

- Level of prefabrication

The most complex solution to his respect is the 'textile-reinforced epoxy resin laminated joint', as it is meant to be formed on site and it requires for the paper-tubes to be fully supported until the joints cure.

The cases 'timber plate puzzle node' and 'timber block node' are mostly prefabricated and can be pre-assembled, but often require post-processing, to optimize the fitting with the tubes, an issue that may increase the effort.

- Workability

The aspect of prefabrication in combination with the effort required for the assembly of the skeletal structure, are hereby considered the most indicative factors in the evaluation process. The case studies with the highest potential for effective assembly are these of the 'fiber-reinforced pulp-composite connector' and 'tolerance adaptive timber plug', that include fully prefabricated elements for easy assembly, as they are fixed together with basic mechanical fixations.

- Maintenance

In principle, joints fixed with fasteners require occasional tightening, as for various reasons, such as displacements in the structural system and changes in the humidity content of the tubes, these are expected to become loose. In this category belong all case studies except for the 'textile-reinforced epoxy resin laminated joint' that involves exclusively adhesive bonding. In this case, damages on the composite shell could be treated with extra layers of resin, perhaps in combination with patches of textile. In case of damage, the 'timber plate puzzle node' would be tricky to fix, because of the timber reinforcements that are laminated on it.

- Reversibility, potential reuse and recyclability

Considering these criteria, the timber block node is the best joining method, as it is fully reversible, reusable and recyclable (downcycling). The same could possibly apply for the 'tolerance adaptive timber plug' that needs to be further developed to become a satisfactory solution.

The 'fiber-reinforced pulp-composite connector' is a fully reversible joint that can be reused when the shells are in good condition. For the moment, the expected end of life scenario is burning.

In a different way, 'timber plate puzzle node' is partly reversible. The timber reinforcements that are laminated on the primary joining system can be removed by force, in order for the node to be recycled. This joint could be reused for a limited amount of times under certain conditions.

The 'textile-reinforced epoxy resin laminated joint' is neither reversible nor reusable or recyclable.

- Critical areas

The case study 'textile-reinforced epoxy resin laminated joint' presents the higher risk, due to the abrupt failure as expressed in the structural tests. Cracks could appear in the composite shell, or delamination and failure could occur on the side of the paper-core. Next to this, in case of damage, reparation is tricky. A possibility that could apply for minor damages is to fill the crack with resin and cover it with a patch. For damages of a bigger scale the structural elements would possibly need to be replaced.

In the case of 'fiber-reinforced pulp-composite connector' splitting between the different layers of the composite could be an issue. The easy assembly method would allow though for fast replacement of the problematic element.

In the case of 'timber plate puzzle node', delamination between the wooden reinforcements and the timber cross could be an issue, but following the compression test (4.2.1.1, fig. 4.26) this scenario does not appear as a high risk.

Table 4.33 Qualitative evaluation of the assembly method planned for the case studies, aspects and practical issues that are important for the phases of implementation and 'end of life'.

Case study, paragraph	4.2.1.1	4.2.1.2	4.2.1.3	4.2.2.1	4.2.2.2
Full Name	Timber plate puzzle node (fixed with screws)	Timber block node (fixed with bolts)	Tolerance adaptive timber plug	Fiber-reinforced pulp-composite connector	Textile-reinforced epoxy resin laminated joint
Joining technique (design outlines)					
Prefabrication level					
Workability					
Maintenance					
Reversibility					
Reuse					
Critical areas					Shell-cracks

Legend

Level of prefabrication

☐ Fully prefabricated connector

▨ Connector pre-assembled or assembled on site

■ On site production (forming) of the connection

Workability

☐ Little effort

▨ Medium effort

■ High effort

Maintenance

☐ Little effort

▨ Medium effort

■ High effort

Reversibility

☐ Dry joint, fully reversible

▨ Semi-reversible or disassembled by force

■ Non-separable joining elements

Reuse

☐ Reusable

▨ Reusable under conditions

■ Non-reusable

Critical areas

☐ Easily replaceable joining elements

▨ For secondary elements of the joint

■ Within the main body of the joint

Prototyping and manufacturing potential

The conclusions drawn hereby summarize the experiences gained through the experimental process of producing prototypes and specimens for further testing and therefore reflect a rather practical approach.

'Textile-reinforced epoxy resin laminated joint' is the easiest joining method for production in terms of required equipment and skills. Still the curing process occupies significant time, a fact that would extend the work on-site. The intermediate joints require more equipment and skills, but can be mostly prefabricated at reasonable costs. 'Timber plate puzzle node' is the easiest solution to realize in terms of technology and skills, however significant time is needed for the lamination of the reinforcements. The components for 'timber block node' can be produced by hand faster than for 'timber plate puzzle node', by someone with the right skillset. In both cases, use of CNC milling technology is the recommendation for easy manufacture. The 'tolerance adaptive timber plug' could develop to the most efficient solution regarding the manufacturing process. In all cases the costs for production are in a medium to low range, except for the 'fiber-reinforced pulp-composite connector' produced with pressure forming (molding).

Table 4.34 Qualitative evaluation of the manufacturing potential for the examined case studies.

Joining system	Technology	Skills	Time		Cost	
			Hand	Machine	Materials	Production
Timber plate puzzle node						
Timber block node						
Tolerance adaptive timber plug						
Textile-reinforced epoxy resin laminated joint				0		
Fiber-reinforced pulp-composite connector			0			

Legend

Technology
- Basic tools
- Special machinery required
- Advanced technology

Skills
- Basic
- Good (material processing)
- Advanced

Time (to manufacture)
- Fast
- Medium
- Slow

Cost (based on the prototyping process)
- Low
- Medium
- High

4.3.2 Structural Performance

As part of the experimental research process, the structural performance of the joints is examined. As stated in the research methodology (paragraph 1.5.3), experimental structural tests and FEM simulations are the instruments used for this purpose.

Regarding the structural tests, due to the lack of standard methods for testing this kind of assemblies, experimental methods were developed. The main priority set was to examine the performance of the joints for the effect of bending. This priority was set following also the susceptibility of paper-tubes for bending. Further than this, the performance of small parts of the assemblies was checked for selected cases, such as the effect of shear forces on fasteners used to fix parts of the joints together. In principle, efforts were made to carry out as many of the experiments with the same boundary conditions (satisfactory quality of specimens, number of repetitions etc.). However, differentiations in the testing process for the case studies exist, mainly because of obstacles met occasionally in producing the specimens for testing. These are clarified within the evaluation process. The series of bending tests performed for the case studies examined are listed in table 4.35. The outcomes of this process are discussed in the following paragraph 'evaluation of final series of experiments'. The focus is on comparing the performance of the different specimens. For this purpose, the limits and failure modes are discussed. Moreover, the interpretation of results is explained, also to calculate some values that represent the performance of the specimens and help to draw comparisons.

For the local FEM analysis, the software used for this purpose is 'ANSYS Workbench Static Structural'. The simulations mainly regard 'maximum principle stress' and the 'total deformation' analysis. The performance of multi-axial nodes is simulated, with the aim to observe complete designs. In this case, the conditions for the applied loads were as decided in paragraph 3.3.2 'Selection of specific context for the case studies'. Next to this, linear configurations of joints, like the ones used for the experimental 4-point bending tests were briefly analysed, to attempt drawing comparisons between these and the failure mechanisms seen in the testing process. The loads applied are, in this case, in correspondence with the respective testing process performed per case. More specifically, the forces applied are equal to $F_{max,Linear}$ as defined by the testing results (further details are addressed in 4.2 'General guidelines', paragraph 'stress analysis of linear configurations'). In this process, three of the case studies were mainly examined, the 'timber plate puzzle node', 'timber block node' and 'textile-reinforced epoxy resin laminated joint'. The output is discussed in the following paragraph 'Local FEM Analysis'.

Based on the previous two milestones the effectiveness of the cases examined is observed. Next to this, the outcomes from the structural tests and the FEM simulations are also discussed, to identify compatibility, differences and possible errors on the different sides and look into how the parallel development of these studies could support the re-search process, to suggest further improvements.

Evaluation of final series of experiments

Hereby the findings from the final series of 4-point bending tests are discussed, based on which the best comparable experiments include the case studies 'timber plate puzzle node', 'timber block node' and 'textile-reinforced epoxy resin laminated joint'. As mentioned in case study 'fiber-reinforced pulp-composite connector', paragraph 4.2.2.1, the manufacturing process imposed limitations for the structural testing. Next to this, case study 'tolerance adaptive timber plug', paragraph 4.2.1.3, was performed only as a proof of concept, towards the end of the research process. The evaluation of the results is adjusted in accordance with these conditions.

Table 4.35 Differentiations in the series of structural tests performed per case study.

Nr.		Full Name	Timber plate puzzle node (fixed with screws)	Timber block node (fixed with bolts)	Tolerance adaptive timber plug	Fiber-reinforced pulp-composite connector	Textile-reinforced epoxy resin laminated joint
		Case study, paragraph	4.2.1.1	4.2.1.2	4.2.1.3	4.2.2.1	4.2.2.2
		Joining technique (design outlines)					
1		Preliminary experimental bending tests (90° configuration, paragraph 6.2)	1 specimen	1 specimen	0	3 specimens (scale 1:2)	1 specimen
2		Preliminary, 4-point bending tests (linear configuration, paragraph 6.2)	1 specimen	1 specimen	0	0	0
3		Final 4-point bending test (set-up shown in fig. 6.3)	Configuration 1, (6 specimens) Configuration 2, (3 specimens) Configuration 3, (3 specimens)	Configuration 1, (3 specimens) Configuration 2, (3 specimens)	1 specimen	0	Configuration 1, (3 specimens) Configuration 2, (3 specimens)
4		Further testing (fasteners as part of the joint)	Yes	Yes	No	Yes	No

Overall, the results gathered by the testing process lie within a wide spectrum, as shown in graph 4.25. The curves for study 'textile-reinforced epoxy resin laminated joint' are the ones that approximate better the behaviour of the paper-tube and create successfully the effect of stiffening in the structural system. Yet, the highest ultimate force is reached by the intermediate joints 'timber plate puzzle node'' and 'timber block node', the curves of which lie close with each other. The 'timber block node' was expected to perform significantly better than the 'timber plate puzzle node', but due to technical errors in the manufacturing of the specimens for 'timber block node' the performance was not optimal. Both these case studies present a weaker axis (configuration -3 for the 'timber plate puzzle node' and -2 for the 'timber block node'), a fact that needs to be considered for the implementation of these in structural systems and also for further optimization.

Regarding the failure modes, two main trends are observed. The paper-tube ruptures close to the end of the joint (where the system suddenly becomes weaker) and following the winding line. When the joint fails, the defects occur close to the center of the specimen, where the moment is maximized, at the weakest link. These trends are commented for the different case studies in figures 4.82 - 4.85.

Graph 4.25 Testing results, force versus displacement average curves.

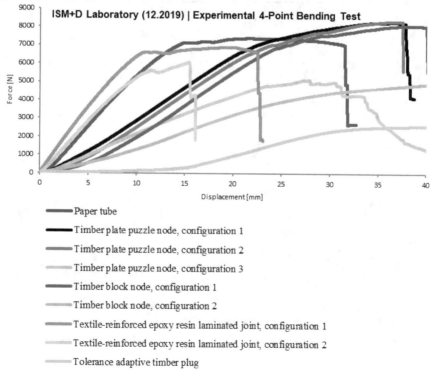

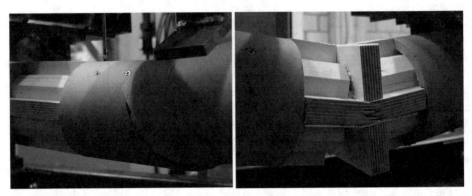

Figure 4.82 'Timber plate puzzle node', failure of paper tubes in the cases of configurations 1 and 2 and failure of timber node in the case of configuration 3, as mentioned in table 4.35.

Figure 4.83 'Timber block node', failure of paper tubes in the case of configuration 1 and plastic deformation observed in the area of the joint, in the case of configuration 2, as mentioned in table 4.35.

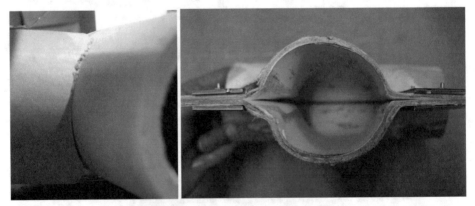

Figure 4.84 (Left) 'tolerance adaptive timber plug', high bending-effect due to the lack of pressure. (Right) 'fiber-reinforced pulp-composite connector, deformation and local delamination within the composite shell.

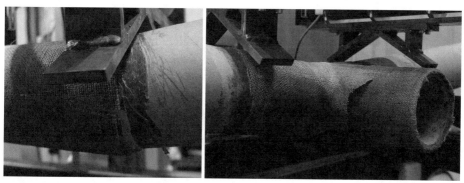

Figure 4.85 'Textile-reinforced epoxy resin laminated joint', (left) In the case of 'configuration 1', for 2 specimens the paper-tube failed and for one specimen cracks occurred also on the composite. (Right) in the case of configuration 2, cracks occurred, every time, on the composite shell.

Representative values

The data obtained by the curves in graph 4.25 are used to extract further values. The start is already made in the respective paragraphs, per case study, where the $F_{max, Linear}$ is set. It is important to clarify that these values represent only the performance of the specimens and do not apply for further assemblies (for example with different length etc.). At first the formulas used for the generation of these values and the assumptions made for this purpose are presented. The results from this process are presented in table 4.37.

(1) Bending moment[1]

The formula used for the calculation is the following (see the results in table 4.36):

$M = l \times (F_{total} \div 2)$, where $l = 0,35m$ and $F = F_{max\ linear}$

(2) Maximum bending stress[2]

This value is only calculated for joining methods for which a constant section can be assumed, in order to calculate the moment of inertia. The sections assumed are presented in fig. 4.86.

$$\sigma_{max} = \frac{Mc}{I}$$

Where:

σ_{max}= the maximum stress un the beam

I= moment of Inertia of cross-sectional area (neutral axis)

c= furthest perpendicular distance from the neutral axis

M= the resultant internal Moment (table 4.37)

---

[1] https://en.wikipedia.org/wiki/Bending_moment
[2] Euler–Bernoulli bending theory (https://en.wikipedia.org/wiki/Bending)

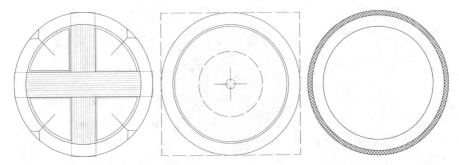

Figure 4.86 Vertical section of the three specimens, from left to right: 'timber plate puzzle node', 'timber block node', 'textile-reinforced epoxy resin laminated joint'.

(3) Bending stiffness [3]

This value is equal to the inclination of the curve (graph 4.25). For the calculation, values that lie within the part of each curve that is assumed as 'linear' are selected.

$$K = \frac{F_1 - F_2}{dl_1 - dl_2}$$

The highest stiffness is reached by the specimens of 'textile-reinforced epoxy-resin laminated joint', a joint based on adhesion. These values surpass the stiffness of the paper-tube itself, even in the case of 'configuration 2', fig. 4.75, 4.2.2.2, where the contact between the tubes that is based on form-fitting creates unfavorable conditions for the performance of the joint.

In the case of plug-in joints, as a result of the tolerances the stiffness is lower. In both cases, 'configuration 1', that represents the primary axis of the timber joint, presents the highest value. The values between 'configurations 1' in 'timber plate puzzle node and 'timber block node' are equal. However, the 'configuration 2' in 'timber block node' demonstrates the lowest stiffness, a result of technical issues that would be different if the joints were produced with CNC milling. Moreover, minimization of the tolerances would improve the contact (friction) and create a positive effect.

Further experiments are required for the 'tolerance adaptive timber plug', with implementation of pressure profiles instead of tensile straps, in order to achieve higher performances.

To use the advantages of each joining method strategically, the different performance expressed in the complementary axes shall be encountered in the process of designing the global structural system.

---

[3] https://en.wikipedia.org/wiki/Stiffness

Table 4.36 Bending stiffness of the specimens tested.[4]

Tested configuration	F1 (kN)	dl1 (mm)	F2 (kN)	dl2 (mm)	Bending stiffness* (kN/mm)
Paper tube	6	11,49	4	7,44	0,494
Timber plate puzzle node, configuration 1	6	18,77	4	13,13	0,355
Timber plate puzzle node, configuration 2	6	19,88	4	14,14	0,348
Timber plate puzzle node, configuration 3	4	18,93	2	9,32	0,208
Timber block node, configuration 1	6	21,47	4	15,87	0,357
Timber block node, configuration 2	3	19	1	8,22	0,186
Textile-reinforced epoxy resin laminated joint, configuration 1	5	7,19	3	4,039	0,635
Textile-reinforced epoxy resin laminated joint, configuration 2	4	7,15	2	3,52	0,551
Tolerance adaptive timber plug	2	27,3	1	20,4	0,145

(4) Spring constant

This value is calculated with reference to the global structural analysis performed in RFEM Dlubal software. There, the opportunity to use this value ($\varphi_y = \sigma_{\varphi y}$), as input within the 'model data', in order to characterize further the behavior of hinged joints is provided, as explained in paragraph 3.2.1, table 3.4. For the global FEM analysis performed prior to the case studies, for the three typical cases (an orthogonal grid, a truss and a dome structure), the joints are assumed as hinged (versions with fixed joints are simulated as well). Then, the implementation of bracing elements minimizes the importance of the spring constant that would otherwise become critical. The reason for this interpretation, within the evaluation, is to maximize the output, based on the results from structural testing and attempt to draw further observations for the performance of the joints.

In principle, for the spring constant, higher values contribute to minimization of moments and shear forces (M_y and V_z). The results are presented in table 4.37. Regarding the applicability of the results on the global structural model, in principle, a small ratio between the length of the joint and the span should apply, so that the length of the joint could then be considered insignificant.

---

[4] The values for F_1, dl_1, F_2 and dl_2 are extracted from the respective graphs.

- Calculation method (spring constant $\sigma_{\varphi y}$):

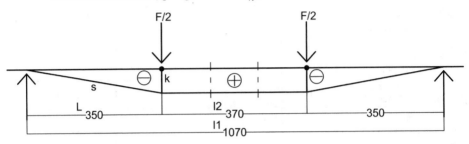

Figure 4.87 Diagrammatic representation of testing set-up.

To define the spring constant $\sigma_{\varphi y}$:

$$\sigma_{\varphi y} = \frac{M}{\Delta \varphi} \ (1)$$

$$M = \frac{F}{2} l_1 \ (2)$$

$$\Delta \varphi = \frac{\delta_{new}}{L} = \frac{\delta_{old} - \delta}{L} \ (3)$$

Below, the vertical deformation δ, is calculated. The following calculations are based on basic static and 'energy methods', particularly on theory regarding the 'principle of virtual forces and unit load method' (Gross, 2011)[5].

$$f = \int \frac{M\bar{M}}{EI} dx \ (4)$$

$$E \times I = constant \ (5)$$

$$f = \frac{1}{EI} \int M\bar{M} dx \ (6)$$

M_i, M_k

$$M_k = \frac{1}{3} sik \ (7)$$

From (6) and (7):

$$\delta = \frac{1}{3} sik \frac{1}{EI} = \frac{1}{3EI} \left(\frac{l_1}{2} - \frac{l_2}{2}\right) \bar{1} \left(\frac{l_1}{2} - \frac{l_2}{2}\right) \frac{F}{2} \left(\frac{l_1}{2} - \frac{l_2}{2}\right) = \frac{F}{6EI} \left(\frac{l_1}{2} - \frac{l_2}{2}\right)^3 \ (8)$$

---

[5] Gross, D. (2011). Engineering Mechanics 2: Springer (p.250, table 6.3).

Comparing the structural performance of the joints examined

Table 4.37 presents the representative values, based on the results from structural testing and the calculations addressed beforehand. On this basis, interesting conclusions and comparisons between the different series of testing and joints can be drawn.

Table 4.37 The representative values calculated for the specimens tested in the final deries of experimental 4-point bending tests.

Collective data							
Type of specimen	$F_{max,Linear}$ (kN)	dl_A support (mm)	dl_B sensor (mm)	Bending stiffness B (kN/mm)	Bending Moment M_{max} (kNmm)	Stress σ_{max} (MPa)	Spring Constant $\sigma_{\varphi y}$ (Nm/°)
Paper Tube	6,1	11,7	0,6	0,493	1.067,5	12,15	-
Timber plate puzzle node, configuration 1	6,5	20,3	0,01	0,355	1.137,5	15,70	495
Timber plate puzzle node, configuration 2	6,3	20,8	0,18	0,348	1.102,5	15,22	457
Timber plate puzzle node, configuration 3	4,3	20,56	-3,55	0,208	752,5	10,39	357
Timber block node, configuration 1	6,5	23	0,85	0,357	1.137,5	15,70	392
Timber block node, configuration 2	3,5	22	-2,6	0,185	612,5	8,46	233
Textile-reinforced epoxy resin laminated joint, configuration 1	5	7	0,2	0,635	875	6,35	5021
Textile-reinforced epoxy resin laminated joint, configuration 2	4,3	7,76	0,58	0,551	752,5	-	-
Tolerance adaptive timber plug	2	27	-0,003	0,145	350	-	-

Legend

Highest performances   Technical issues suspected for low performances

Limited data (only one specimen was tested)

- Testing series that demonstrated the highest $F_{max,Linear}$ and stress σ_{max}

Configurations 1 for the 'timber plate puzzle' and 'timber block' nodes present the highest $F_{max,\ Linear}$ and in continuation the highest stress σ_{max}. Based on the findings, the joining method 'timber plate puzzle node' presents the most reliable combination of values that lie in the higher range, following the properties of bending stiffness and strength. From a practical perspective it could be said that the form-fitting between the cross and the tubes is responsible for the satisfactory structural performance that exceeds this of case study 'timber block node'.

- Testing series that demonstrated the highest bending stiffness

Configuration 1 in 'textile reinforced epoxy resin laminated joint' presents the highest stiffness and spring constant with remarkable difference from the rest of the series. However, the abrupt failure of the specimens, lacking warning signs, is an issue that creates necessity for further development of this joining method, to improve its reliability.

- Spring constant

Implementation of the values obtained for the spring constant on the global FEM analysis model for typical case 1, orthogonal grid, indicates that due to the presence of tension elements used for bracing, all groups of values from table 4.37 are in principle acceptable for a stable structure under certain conditions. These include the condition of the idealized beam elements and the exclusion of further imperfections, following the environment of the software and the input provided as stated in the introduction of chapter 3.2 'global FEM analysis'. As expected, significant differences are observed in the deformation $U_{t.max}$, within the range of values presented in table 3.6, paragraph 3.2.2.

Local FEM Analysis

Comparing the performance of the multi-axial nodes as analysed in ANSYS software, the maximum total deformation is on the same level for case studies 'timber plate puzzle node' and 'timber block node' and lower for 'textile-reinforced epoxy resin laminated joint'. The resulting stresses create critical conditions in the following cases, as indicated in the output listed in table 4.38 and also the exports presented in fig. 4.88:

- Case study 'timber plate puzzle node': on the timber reinforcements, at the area of the screws (4.2.1.1, fig. 4.28) and at the bottom corners of the timber plates.

- Case study 'timber block node': in the direction of 'configuration 2', within the timber-block, at the area around the steel thread and the connection point between the timber-blocks (4.2.1.2, fig. 4.47 and 4.48). Based on these outcomes, the form-locking joint needs to be optimized to enhance stability. The enlargement of the joint could also be an option, also in combination with paper-tubes of higher diameter.

- Case study 'textile-reinforced epoxy resin laminated joint': along the seams between the paper-tubes and especially in the contact area between the composite-shell and the fixation of the steel plates for the bracing elements (connected with adhesion) (4.2.2.2, fig. 4.78).

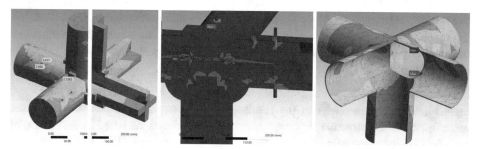

Figure 4.88 Analysis of multi-axial nodes in ANSYS Workbench, Static structural. From left to right: 'timber plate puzzle node', 'timber block node', 'textile-reinforced epoxy resin laminated joint'.

Table 4.38 Output from FEM simulations performed in ANSYS Workbench (Static Structural).

Multi-axial models			
Joining technique	Maximum principal stress (main elements of the joint) (MPa)		Maximum total deformation (mm)
Timber plate puzzle node	Steel	187.7 (screw)	7,79
	Plywood	2.1	
	Douglasie timber	18.3	
	Paper tube	2.84	
Timber block node	Steel	263 (Bolt)	7,13
	Natural timber	30,7	
	Paper tube	6,46 (Bolt area)	
Textile-reinforced epoxy resin laminated joint	Steel	144	4,96
	Composite	5,72	
	Paper tube	2,9	

Legend

The stresses exceed the capacity of the material

Comparing the 'local FEM analysis' with 'experimental testing'

Regarding the comparison of the testing process with the numerical analysis performed for the linear configurations of joining techniques, overall, following the individual reports, the resulting stresses at the support and the failure mechanisms identified in both cases are comparable. In some cases of disagreement between these outcomes, the processes of structural testing and FEM analysis complement each other. Two relevant examples for this are provided below.

In the first example FEM analysis reflects the potential of a joining method that was not well represented by the testing process due to technical issues. In the case study 'timber block node', configuration 2 (see fig. 4.90), the deformation estimated by the software is significantly lower than in the actual test. In fact, if the joint would be manufactured with higher precision and smaller tolerances, then it is expected that this gap would be much smaller.

In case study 'timber plate puzzle node', configuration 3, FEM analysis (fig. 4.89 - and fig. 4.34, 4.2.1.1) indicated that the resulting stresses are more critical for the tubes, when in fact, in the structural testing process all three specimens failed in the middle of the joint. There the veneers of the multiplex separated and cracked. However, the FEM analysis was equipped with a simplified material model.

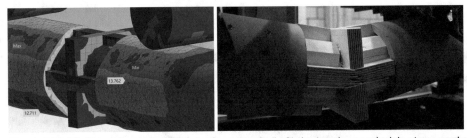

Figure 4.89 'Timber plate puzzle node', configuration 3. (Left) the 'maximum principle stress analysis' indicates highest stresses and failure of the paper-tubes, caused however by the weakness of the joint. (Right) During the testing process for the same configuration, the joint failed systematically. The simplifications assumed for the FEM model lead to this difference.

Figure 4.90 'Timber block node', configuration 2, (left) maximum principle stress analysis, (right) elastic-plastic deformation for the specimen (steel bar). The 'ANSYS' simulation indicated that the timber blocks would be in danger of damage, on the inner edge of the insert. However, during the structural tests the only obvious damage observed systematically (in all three repetitions) was on the steel bar that bent.

Summary

Overall the research process, as a combination of designing (detailing and assembly method), prototyping, performing structural tests and simulating the structural performance of the different joining methods, provided a comprehensive approach and a helpful tool-set to proceed with.

Table 4.39 Global observations on the case studies presented in chapter 4.2.

Joining system	Design flexibility	Easiness in Assembly	Efficiency in Production	Structural integrity	Dura-bility	Research documenta-tion
Timber plate puzzle node	(+) Structural					
Timber block node	(+) Structural		Optimization required			
Tolerance adaptive timber plug	(+) Assembly	Optimization required	0*	0*	0*	
Textile-reinforced epoxy resin laminated joint'	(+) Production, structural		Optimization required			
Fiber-reinforced pulp-composite connector	(+) Assembly		Optimization required	Optimiza-tion required		

Legend

High ▢ Medium ▨ Poor ▬ *0= not evaluated

Design flexibility

As described in the observations made above, the case studies examined present good grounds for implementation in different designs. Still their different qualities make them more or less suitable for different designs. Thus, to conclude on the most suitable joining technique to implement in a global structural system, the requirements for structural performance and the possibilities for manufacturing need to be considered altogether.

Prototyping

Due to the experimental character of the case studies, significant effort was required to produce the specimens. Still, for all joining methods studied, the technological means that are needed to proceed with further production are widely available. Hence, there is potential to optimize the individual manufacturing processes and improve the issues of imperfections that are mentioned for some cases. The processes of pressure forming (used for the 'fiber-reinforced pulp-composite connector') and 5d-milling (alternative automated process for the 'timber block node') are the two examples that depend on more advanced machinery.

Assembly method

To conclude on the efficiency of the assembly methods, the concept of the 'fiber-reinforced pulp-composite connector' presents clear advantages. This result is a combination of different parameters. Satisfactory precision is achieved by the pressure-forming that leads to optimal form-fitting and the thin shells form a steady lap joint that is easy to penetrate, in contrast to other methods, in order to apply the fasteners (bolts). At the same time further development of the prototypes and investigation of the structural performance is necessary, following the reported outcomes.

Structural integrity (structural testing and local FEM analysis)

Within the process of structural testing, in the majority of the tests, the failure occurred on the paper tube, on the side of the joint. This happened both due to the placement of the force on top of that area and also the different behavior of the system within and outside the area of the joint. Regarding the aspect of structural integrity, based on the results from structural testing, the specimens for study 'timber plate puzzle node' present overall the highest bending strength, whereas 'textile-reinforced epoxy resin laminated joint' the highest bending stiffness. As it has been stated already, technical errors in the manufacturing process may have worsened the performance of the specimens for the study 'timber block node' which were expected to present the highest bending strength. Considering the different qualities presented by the cases studies, combination of the differrent joining methods to build a skeletal structure would be interesting.

The local FEM Analysis highlights the issues that rise with the integration of mechanical fixations in the assemblies and the benefits of a joint that relies on adhesion.

A global conclusion from both the structural testing and the FEM analysis is that paper-tubes with higher structural performance are required. Thus, implementation of the studied joining methods on tubes with different geometry would be interesting.

Research documentation

For this research, precedents related to the study 'timber plate puzzle node' provided a first base for discussion. The experiments performed produced new insights on the functionality of this joining method and also enriched the spectrum of possibilities.

5 Global observations

Keywords: paper-based materials, joining techniques, state of the art, paper tubes, multi-axial joints, beam structures, hybrid joints

In this chapter the findings from this research are summarized. Prior to this, the research approach and process are repeated here, to emphasize on the storyline in which the following discussions in paragraphs 5.1 - 5.4 belong. The subject investigated is this of joining techniques for assemblies built with paper-based components. Overall the research approach lies on a medium between architecture and engineering, in terms of the equipped research tools. The greatest challenge is related to the novel character of the topic and the minimal input for the aspect of joining techniques, as paper-based components are not established building materials. The difference from daily applications to more robust building applications is great when considering the requirements regarding the performance on various levels, such as structural aspects and durability. The aforementioned issues are reflected by the research questions, stated in chapter 1 'Research methodology', that basically set the requests for designing the global picture for joining techniques in this field, outlining the future potential, to foresee opportunities and weaknesses and then experiment with selected solutions that lie closer to full-scale construction.

Therefore, the scope of the research on joining techniques is rather broad in the start of the thesis, to address all the parameters that shape the 'state of the art' and its future potential development, including the relation with conventional building materials. This part is viewed as an opportunity to present paper-based products in a new frame and discuss the applicability of joining techniques based on precedents, sketched details and draft experiments. The outcomes are discussed in paragraph 2.4 'Global overview of joining techniques for paper structures'. Afterwards a specific area is targeted, to examine in detail a selection of joints, as case studies, for the construction typology of multi-axial joints for beam structures composed of paper-tubes, a direction with highly valued potential, as explained in paragraphs 2.1.1 'Collection of reference projects and studies', 2.2.4.1 'Joining techniques for paper-based tubes', the epilogue of paragraph 2.4 and 3.1.1 'Construction typologies'. To comprehend the structural issues to be dealt with, three typical cases of beam-structures are analysed in paragraph 3.2 'Global FEM analysis'. In chapter 4 'Case studies', the experimental research on a variety of joints is presented, in the two categories of plug-in and sleeve joints. The following aspects are examined systematically: detailing, assembly techniques, prototyping and production techniques, structural analysis performed with structural tests and FEM analysis. The

E. Kanli, *Experimental Investigations on Joining Techniques for Paper Structures*, Mechanik, Werkstoffe und Konstruktion im Bauwesen 60, https://doi.org/10.1007/978-3-658-34501-3_5

outcomes of this process are presented hereby, with extra focus on the structural performance and the results from the bending tests.

Structure of the research outcomes

In this chapter the research findings are compiled, in relation with the research questions and objectives, as these are stated in chapter 1 'Research methodology'. In principle, the research process followed, as it is addressed in paragraph 1.4 'Process and milestones', is comprehensively developed in chapters 2 - 4. The role of this chapter is to clarify how the elements of the research process and the derived findings per case contribute to the required answers. For this purpose, in paragraph 5.1, for each of the research questions, the findings are summarized and the location of all relevant information in this dissertation is identified. In continuation, in paragraph 5.2 'Outlook', topics for optimization and further research are addressed. The chapter is completed with a reference to the teaching process (5.3), as part of the research activities and an epilogue (5.4), where a statement on the outcomes of this research and further indications deriving from these is made.

5.1 Research questions and findings

In this chapter, the research questions followed by the efforts made to deliver answers are presented together. For this reason, per question, the main steps followed in the research process are stated and the outcomes are discussed.

Research question 1

What is the 'state of the art' of joining principles for experimental structures built primarily with paper-based materials?

    a.   Which are the main categories of joining principles that can be identified within the 'state of the art' and what are the main pros and cons for implementation in construction?

    b.   How could the 'state of the art' of joining principles develop in the future, considering further potential solutions?

The process followed to identify the 'state of the art' is presented step by step in chapter 2. The review of joining techniques is divided in two main sections in paragraphs 2.2 'Joining techniques for paper-based assemblies' and 2.3 'A review of joining techniques for common construction materials' that come to satisfy the sub-questions a and b. The global 'state of the art' is presented in paragraph 2.4 'Global overview of joining techniques for paper structures'. The same sequence is kept below as well.

- Research question 1 a

Focusing on assemblies of paper-based components, in chapter 2.2 the state of the art is analysed in the three main sections of 'basic joining techniques', 'techniques specialized on lightweight paper-based boards' and 'techniques specialized on paperboard profiles'.

This distinction of the joining principles in categories is made based on the type of paper product, defined by the material-composition (thin layer or rigid board), the geometrical characteristics (flat board or profile, range of thickness) and the application (used exclusively in daily products or with precedents in construction related applications etc.).

A further distinction between the joining principles presented per category is made, based on the principle joining method expressed per case, in the three clusters of material-closure, form-locking and force-closure. This categorization method is inspired by the design catalog for fixed connections, as presented in the guidelines for engineers about the 'methodical selection of solid connections', as an assisting tool for the engineering process (VDI, 2004, p. 26, catalog 1). Joining techniques that use more than one of these methods are identified as hybrid. This approach is used throughout the whole review process and also in the case studies, presented in chapter 4, to describe the functionality of the hybrid joints investigated.

In fig. 5.1 the sub-overviews for the three categories, under which the joining principles for paper-based assemblies are identified, are presented once again, to support the discussion. In continuation, the findings in this area are summarized.

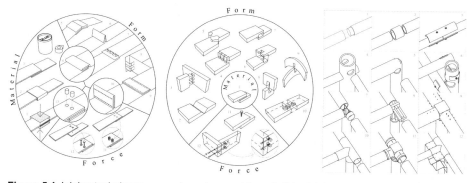

Figure 5.1 Joining techniques, paper-based assemblies - (left to right) table 2.4, table 2.5, table 2.6

In the category 'basic joining techniques' fundamental solutions such as adhesion, stapling, stitching, interlocking (creased edges and flaps) and snap connectors are examined. Lamination is the most important joining technique, best explored by the paper industry, where adhesives such as starch, dextrin, water-glass, PU and PVA glue are used in differ-rent occasions. From the adhesives mentioned above, PVA glue is commonly used in timber connections and therefore a variety of products is available. Draft tensile tests on laminated lap-joints formed with kraft paperboard and 'Ponal' glue (PVA), performed in this research, showed failure on the side of the paperboard and indicate potential relation between the fiber orientation in the paperboard and the behavior of the laminate at ultimate limit state (paragraph 2.2.2, graphs 2.1 and 2.2). Starch-based adhesives are known as a more ecologically friendly solution, but often present lower strength. The joining techniques based on form-locking and force-closure mentioned above present in general limited structural performances.

In the category of 'specialized solutions for lightweight paper-based boards', 3d-interlock (puzzle) assemblies are often used to create lightweight designs. In the area of force-closure modern fasteners, such as screws with wide spiral that anchor better in the board and cross-dowels are used to fix honeycomb boards together, mostly for assemblies related to packaging products and furniture. Assemblies that lie closer to the field of construction regard hybrid solutions. Examples are found in realized structures that present however fundamentally different techniques, depending on the global structural system. When u-sing paper-based boards in construction, the penetration of the boards is in principle avoided. Methods that involve adhesion or applying moderate pressure are more often preferred. An example for this is the product 'Wikkelhouse' – see p. 47- (Fiction factory). Another example is shown in Nemunoki Children's Museum' (McQuaid, 2003) where pressure profiles bolted on steel rings are used to assemble a lamella structure for the flat roof that is made of honeycomb board, without penetrating the boards (see fig. 2.15). A different approach is shown in the images from the 'Apeldoorn Temporary theatre' (fig. 2.20), where a lamella structure is assembled without the involvement of any fasteners, but multi-layered laminated paperboard lamellas, connected in terms of form-locking at the nodes that are secured with pressure ties, to prevent deformations due to shearing.

In the category of paper-based profiles there is a variety of precedents for beam structures composed of paper-based tubes, where the elaboration of the joints is in the main focus and a number of different assembly methods is demonstrated. The simplest examples, categorized as basic techniques, involve simple form-locking and implementation of direct mechanical fixations (fig. 2.26). More durable solutions regard hybrid techniques which in their majority include different types of nodes, made of timber (fig. 2.29) or steel (fig. 2.39 - 2.41). The first ones present the advantages of simplicity and practicality, but are not as effective in a structural manner without the implementation of reinforcements. The integration of adhesives could improve the performance of the joints, however with drawbacks to the ecological factor. Most of the steel intermediate connectors reviewed are fully functional, especially because they are mostly combined with pre-stressing the paper-tubes. However, such joints fit better in highly-durable structures rather than temporary ones. On this matter, joints designed with wood-products are observed as a more suitable solution that could be realized with more accessible technologies, comparing to the manufacturing of steel nodes. The category of joining techniques for paper-tubes is the main focus in the following paragraphs.

- Research question 1b

Here the aim is to find potential improvements for the joining techniques identified in the previous section and also to consider further joining techniques that present potential. In this spirit, in the second part of the review, the potential of joining techniques that apply for the building materials of timber, bamboo, textile, composite and steel products for paper-based assemblies is discussed. The outcomes from the review process are briefly summarized per material category in paragraph 2.4.1. Overall, the fields of timber and bamboo structure provide the richest input for this research (see fig. 5.2).

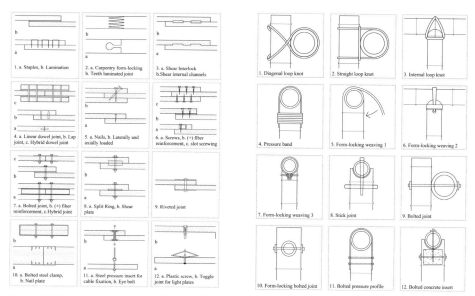

Figure 5.2 Joining techniques in timber (table 2.7 – original) and bamboo (table 2.8 – original)

In timber structures, practices on lamination are particularly useful, also in relation with CLT construction. In the case of screws and bolts, regarding the failure modes as an issue that is still unexplored and problematic for paper-based components, the guidelines provided in EN1995-1-1: Section 8 - Connections (Leijten) are a useful base, despite foreseeable differences. On the other hand, joining techniques that rely on form-locking (carpentry joints) present significant issues if applied on paper-based products, depending on the matrix - layering - of the assembly.

Bamboo structures present assemblies closely related to paper-tube structures. Joints that combine form-locking and gluing could be compatible. However, cut-outs on the tubes could weaken the profiles significantly. Examples of intermediated connectors made of wood (fig. 2.68), steel (fig. 2.69 – 2.72) and concrete (fig. 2.73) could be transferable. Designs that offer increased contact surface between the profile and the joint present better potential.

In the case of textiles, methods for local reinforcement of the material at the areas of the joints, either with integration of rigid plates or patches, to improve the performance of mechanical fixations for instance, prevent damage and increase durability are only some ideas in this direction. In a similar way composites present more advanced methods, comparing to textiles, for the embedment of fasteners in components, within the manufacturing process. Moreover, studies that examine the integration of fiber-reinforcement at the areas of the joints to prevent material damage from punctual joints present interest. Steel inspires for innovative concepts of techniques comparable to welding and related to couching, a method used to embed layers of paper together.

- Global overview of joining techniques

Table 5.1 Overview of Joining Techniques – Map of influences and solutions (originally 2.12)

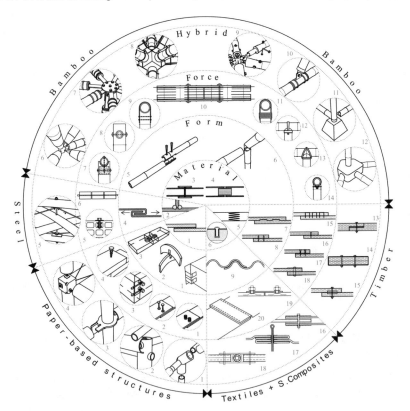

In paragraph 2.4.2, an overview is presented that combines the outcomes from both the review of joining techniques for paper-based assemblies and the techniques identified as influences from conventional building materials, with the aim to sketch the greater potential for the development of further studies and research. A qualitative evaluation of all included techniques is expressed in the accompanying tables 2.13 - 2.16, as a summary of this process. There, all techniques are named, with reference to the numbers viewed in table 5.1 (originally 2.12). The criteria considered are the transferability of forces (pressure, tension, shear and moment), the level of technology required for manufacturing, critical aspects for the implementation and potential areas of application for the technique.

- Research question 2

Considering that the product of paper tubes is the most highly applicable paper-based component in construction so far, particularly for load-bearing elements, what are the most prevalent joining techniques implemented and why?

> *a. As part of exploring the potential implementation of paper-based products in construction, how can the preference to paper-tubes comparing to other products be explained and what other types of structures that make use of other paper products can be identified?*

> *b. Exploring the potential of this construction typology, what structural configurations and applications are suitable, based on the characteristics of the paper-tube? Next to this, which joining methods are suitable (also in combination with research question 1)?*

As the focus of the research shifts towards a specific area of interest, this question raises three issues, with focus on the component of paper-tubes, the most preferred joining techniques and the potential structural systems of interest. The issues addressed here are examined in more than one steps in this research. These steps are identified first, prior to the discussion of the findings.

The most prevalent joining techniques observed in structures composed of paper-tubes are reviewed in paragraph 2.2.4.1 'Joining techniques for paper-based tubes' and key-observations are made also in the epilogue of 2.4.

The applicability of paper-based products in design and construction projects is discussed in paragraph 2, together with the 'state of the art'. The comparison regarding the applicability of paper-tubes versus other products and especially lightweight paper-based boards, for such purposes, is addressed in paragraph 2.1.1 'Collection of reference projects and studies', where a selection of projects for both cases is presented and the advantages and disadvantages, based on the observed building methods and previous experiences are summarized. In addition to this, the subject of construction typologies observed throughout this research is briefly discussed in paragraph 3.1.1, where an introduction to paper structures is made prior to the analysis of beam-structures composed of paper-tubes. In continuation, a specific type of paper-tube is selected that is presented in paragraph 3.1.2, for the reason of developing realistic boundary conditions and input for the research that follows. In paragraph 3.1.3 'Construction outlines and typical cases' three configurations of structural systems are examined and insightful observations are made, based on the global structural analysis performed.

- Findings on the most prevalent joining techniques for paper-tube structures

Multi-axial joints are examined closely when analysing the 'state of art'. A wide range of joining techniques are reported. Influences from bamboo structures are obvious in examples of form-locking connections, as well as from timber and steel structures in examples of intermediate connectors. The most important findings are categorized as following:

Techniques that are based on form-locking, include both simple and hybrid solutions where reinforcements may apply, such as adhesives, ties or fasteners:

1. Simple lap joints, fixed with bolts and secured with ties (fig. 2.26)

2. Form-locking joints combined with adhesion and ties (fig. 2.25)

These techniques are used for short-term installations with low structural requirements, due to advantages for fast production and erection. When implemented in structures of bigger size, as at the 'Japan Pavilion' (McQuaid, 2003), fig. 2.28, then additional structural support is provided.

Intermediate connectors (techniques 3 – 7, below) are used when higher strength and durability are required. These can be divided in plug-in and sleeve joints. Plug-in connectors are distinguished according to the material (steel, timber etc.) and the assembly method:

3. Plug-in connectors made of interlocked timber plates (fig. 2.29 – 2.32)

4. Plug-in connectors made of massive wooden blocks (fig. 2.33 – 2.36)

Overall techniques 1 – 4, presented above, are highly preferred by project-teams that undertake the construction process as well, further than the design. In this context, connectors made of timber plates (3) present advantages for the production of the connection points and thus have often been used for emergency shelters. On the other hand, connectors made of wooden blocks (4) show, in principle, structural and aesthetical advantages, but the manufacturing process is more demanding. ·

5. Steel plug-in connectors (insert mostly made simply of steel pipes or blocks of wood) (fig. 2.38, 2.39, 2.41 – 2.45)

Steel connectors offer the highest structural performances as well as aesthetics and therefore are used for long-term installations. This technique (5) includes various alternatives for the assembly method. Often the joining elements are inserted in the tubes with hydraulic pressure and are either fixed directly with the tubes, with screws, or they are fixed with steel bars, placed in the center of each tube, that are used for pre-stressing (fig. 2.38). The technique of prestressing the paper-tubes elevates the stability of the structure, as the capacity for axial compression is in the main advantages of this component which make it so popular for structural applications, comparing to other paper-based products. Still, due to the fundamental differences between paper-tubes and steel, especially regarding the lifespan and cost of the materials, their combination for construction purposes creates contradictions that cannot be overlooked.

6. Sleeve connectors made of steel (fig. 2.40) form a shell around the adjacent beams and the assembly is secured with fasteners.

Sleeve connectors present aesthetical advantages, but also design principles that are more complex to manufacture. Such an example is presented in project 'Boathouse', FRA, (Miyake, 2009, p. 74). Filling the endings of the tubes, in this case with wooden blocks, to prevent them from deforming, as a result of the pressure applied on them, is a known practice that is also observed in similar assemblies from bamboo structures (only then concrete is used instead).

Overall, techniques that use primarily renewable materials to form connection-points gained more focus, due to the global aims of the research regarding sustainability.

- Research question 2a

The findings presented in the 'state of the art', in 2.1, indicate that joining techniques and assemblies for construction purposes reach a higher level of detail and elaboration in the case of paper-tubes, comparing to other paper-based products. The opportunity to use the component with minimum post-processing, but also the easy accessibility and low costs are some of the parameters that make this product popular. The manufacturing process that allows for the production of profiles with different geometrical characteristics and qualities in terms of strength (based on the paper-sheets used and the wall thickness) creates flexible conditions for the design process. Additionally, the round shaped profiles present structural advantages comparing to profiles with sharp corners and in combination with the solid wall, the structural capacity, particularly for axial compression force, reaches promising performances. Specific information about the structural performance of the paper-tubes used in this research is provided in paragraph 3.1.2.

On the other hand, multi-layered paper-based composite boards require the development of modern equipment to achieve effective results in the manufacturing of components for construction purposes and complex studies to examine the behavior of the finished components.

On these grounds, for the moment, skeletal structures composed of paper-tubes demonstrate advantages for the development of load bearing assemblies for a wide spectrum of applications in temporary structures, comparing to other paper-based products.

- Research question 2b

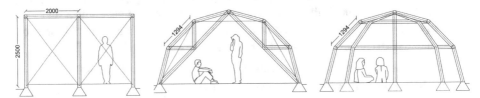

Figure 5.3 (From left to right) orthogonal grid (3.7), truss structure (3.8), dome structure (3.9)

The typical cases of structural configurations examined in this research refer to three different geometries of beam structures, an orthogonal grid, an arch-shaped truss and a dome, as potential environments for the implementation of the joining techniques developed in chapter 4. These configurations are met in reference projects and demonstrate fundamental structural systems that create interesting alterations for the design of the joints.

For these cases, global FEM analysis is performed in software 'RFEM, Dlubal'. The aim is to create better understanding over the structural issues for the members and the nodes, by looking at the distribution of loads and also to select a scenario for the elaboration of case studies and the input for the local FEM analysis (chapter 4).

The paper-tube component selected for the experimental research process has inner diameter 100 mm and wall-thickness of 10mm. The mechanical behaviour assumed based on prior studies is described in table 3.2. The boundary conditions and assumptions made for the build-up of the simulations are described in detail in the respective paragraph (3.2.1). The simulations are performed both for the cases of hinged and fixed joints. In this spirit, the resulting forces, the total deformations and the stability of the systems are observed. The stability is examined with the use of an add-on module (RSBUCK) that is embedded in the software.

In all three cases the most critical issue is related to the bending of the profiles. Therefore, the design of a structure with a flat roof would be the most challenging case, comparing to an arch or a dome, as indicated by the stability factor as well. A simple way to minimize the deformation would be to densify the grid, especially in the case of a structure with an orthogonal grid. An optimization could regard profiles with better structural performance. Such a step would require input for the structural behaviour of the profile and therefore structural tests to create the necessary input.

Focusing on the functionality of the joints, the global structural analysis comes to underline some important aspects. One of them is to transfer successfully the normal forces and keep the tubes under compression, to minimize other effects. Thereby, from the design perspective, increased contact surface is one of the challenges to overcome, due to the round hollowed geometry of the tubes. Next to this, it is crucial for the joints to absorb the lateral forces and moments that occur on all sides, to minimize the effect of bending in the tubes. Considering the effects of wind-loads that are expected to cause shearing between the structural elements and consequently between the elements of the joints, one more priority is added in the picture.

- Research question 3

Which joining techniques show the best potential for small scale tubular structures, to be used as a temporary living space, considering the aspects of design, assembly, stability, materialization and production?

a. *What possibilities shall be further explored, as case studies, in accordance with the findings following research questions 1 and 2?*

b. *What are the greatest difficulties in designing suitable joints for tubular structures, especially considering the requirements for structural integrity? Continuously, what are the most critical aspects for the structural integrity of the joints for the different cases examined?*

c. *Classify the pros and cons identified for the case studies elaborated within this research, considering the main aspects addressed in question 3 and also other aspects that might appear to be important as indicated by the research process.*

The answers are mostly part of chapter 4 'Case studies', where the investigations on selected joining techniques are elaborated. The main research question is answered in paragraph 4.3 'Evaluation of case studies'. The experiences from the development of case studies are reported extensively in 4.2. For sub-question 3a interesting findings are presented in paragraph 4.1 'Design concepts and selection of case studies'. For sub-questions 3b and c insightful observations are provided altogether in paragraph 4.3. Below the relevant findings are summarized.

- Research question 3a

A showcase of design concepts that include techniques based on form-locking, plug-in and sleeve joints is presented in table 4.1, as possible directions for the following research investigations. A qualitative evaluation leads to the selection of concepts for further development. The main aspects considered are these of 'transferability of forces', production, cost, assembly/ disassembly, durability and recyclability.

Based on the 'state of the art' presented in chapter 2, plug-in connectors, made of wood, are presented in reference projects. Overall, plug-in joints help to increase the contact surface and thus optimize the distribution of stresses between the adjacent beams. The use of materials with wood origins is particularly attractive as a solution to complete a paper-based structure. The easiness in production, as a result of simple forming methods is another benefit. Still, there is very limited information regarding technical aspects and particularly the structural performance of these types of joints. On these grounds, further investigation of these concepts is considered particularly interesting.

Simplifying the design and production further and allowing for minor errors in the assembly are interesting points for further investigation.

In a similar way, sleeve joints also help to optimize the transferability of forces, on a form-locking based beam to beam connection. However, the most known examples regard metal connectors. The concept to develop thin-walled joints realized with bio-based materials is an interesting alternative.

On these grounds the selected case studies are presented in fig. 5.4.

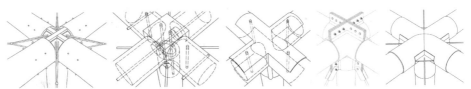

Figure 5.4 (From left to right) Timber plate puzzle node (4.2.1.1), Timber block node (4.2.1.2), Tolerance adaptive timber plug (4.2.1.3), Fiber-reinforced pulp-composite connector (4.2.2.1), Textile-reinforced epoxy resin laminated joint (4.2.2.2)

- Research question 3b

Below the greatest challenges in the elaboration of case studies are summarized.

A global design problem with significant impact on structural performance, especially the effects from bending and torsion, regards differences of the assembly method per direction of the multi-axial joint. The aim is to achieve comparable performances in all axes. In all case studies such issues are confirmed by the results from structural testing. These were already suspected, but the results from testing helped to quantify the problem for the case of bending and also observe differences in the failing mechanisms. This is a disadvantage comparing to the plug-in steel nodes presented in the 'state of the art'.

A major challenge in the category of 'plug-in' joints is to maximize the contact surface between the joint and the tube in order to transfer forces more effectively. Difficulties related to imperfections in the manufacturing process and tolerances between the different elements of the assembly prevent optimal structural performance. Moreover, the practice of prestressing the tubes is very beneficial but tricky to implement with wood-based connectors. Still it is compatible with case study 'timber block node'.

Further, in the case of 'plug-in joints', identifying an acceptable tolerance that would allow for easy assembly and at the same time ensure sufficient contact between the elements was a trial and error problem. Based on the prototyping experiences a tolerance between 1 - 2.5 mm is considered effective. Next to this, it should be noted that differences of ± 5 mm in the inner diameter of a tube is possible.

Once again for the category of plug-in joints, fixing the connector in the tubes is an issue. As part of the experiments, both screws and bolts were tried. Both solutions were found to be effective following the bending and compression tests. The greater concerns are related to the performance under a combination of loads and even more to the long-term performance, also under the influence of alternating climate conditions. The local FEM analysis performed for case-studies 'timber plate puzzle node' and 'timber block node', in 'ANSYS' software, indicates critical conditions at the areas of mechanical fixations with the peak stresses occurring there that could cause damage on the tube and the wooden joining elements.

Moreover, the effective integration of tensile elements as bracing to stiffen the structure is crucial. Local reinforcement for the fixation of these on the joints is essential. The study 'textile-reinforced epoxy resin laminated joint' presents the most challenging case for the integration of such fixations and perhaps the integration of a metal-base, for this purpose, in the manufacturing process shall be considered. In general, as part of future research, pull-out tests to identify the performance of screwed or laminated connections for tension would be an interesting point to examine.

As the main focus of the structural experiments was on examining the performance of assemblies when subjected to bending, designing joints that reinforce the adjacent beams is a great challenge. Only the study 'textile-reinforced epoxy resin laminated joint' shows potential for achieving this goal, when the design of the layers would be optimized. However, the issues of durability and recyclability are important downsides in this case.

In the case of 'fiber-reinforced pulp-composite connector', the optimization of the matrix with higher shell-thickness is crucial, in combination with optimization of the joining technique between the adjacent shells, to stiffen the connection.

- Research question 3c

The pros and cons and comparisons between the case studies examined, are extensively reported in paragraph 4.3 'Evaluation of case studies'. Here a selection of considerations is addressed, to highlight some of the courses learned through the research process.

- Design flexibility

With respect to the aspect of 'design flexibility', the case studies examined present good grounds for implementation in different configurations. Still, the relation between design and structural integrity or manufacturing process could indicate one solution or another as more suitable for a certain design or construction project. This point becomes clearer with the following discussions on both aspects.

- Assembly method

Considering the level of prefabrication and workability, the case study that requires the most effort is 'textile-reinforced epoxy resin laminated joint'. The joints 'timber plate puzzle node' and 'timber block node' are mostly prefabricated but often post-processing of the joints is required, to optimize the fitting, before the structure could be assembled. Case studies 'fiber-reinforced pulp-composite connector' and 'tolerance adaptive timber plug', that include fully prefabricated elements, easy to fix present the highest potential for effective assembly. Still, both cases require optimization to become applicable.

- Prototyping and manufacturing potential

The tools required to hand-craft the joints examined are easily accessible for most cases, as well as some widely available automated processes, such as 2d or 3d CNC milling. The manufacturing processes of high-pressure forming, used for the 'fiber-reinforced pulp-composite connector' and 5d-milling, the alternative automated process for the 'timber block node', are the two examples that depend on more advanced machinery.

- Structural integrity

In most of the 4-point bending tests, the failure occurred on the paper-tube. Cracks developed on the bottom side (tension), opposite from one of the loading points, close to one of the ends of the joint, due to the different behaviour of the specimen within and outside the area of the joint. According to the results from structural testing, the specimens for study 'timber plate puzzle node' present overall the highest bending strength, whereas 'textile-reinforced epoxy resin laminated joint' the highest bending stiffness. Technical errors in the manufacturing process seem to have worsened the performance of the specimens for the study 'timber block node' which were expected to present the highest bending strength. The local FEM Analysis highlights the issues that rise with the integration of mechanical fixations in the assemblies and the benefits of a joint that relies on adhesion. A global conclusion from both the structural testing and the FEM analysis is that paper-tubes with higher structural performance are required.

5.2 Reflection

5.2.1 Outlook

As part of the research process intensive prototyping and efforts to analyse the structural performance of the joints took place. Based on these experiences, points of interest for future development are summarized below.

Joining techniques

First of all, the design concepts presented in table 4.1 that were not further elaborated (such as the sleeve joint of concept 10), could be the starting point for further research.

Further research on the case studies elaborated could regard:

- Optimization of the manufacturing methods and further structural testing of specimens ('timber block node', 'tolerance adaptive timber plug' and 'fiber-reinforced pulp-composite connector')

- Investigations on bio-based binders that could replace resin in case study 'Textile-reinforced epoxy resin laminated joint'.

- Axial compression tests of assemblies, similar to the ones presented in pages 212 and 242, with the focus on the performance of fasteners.

- Observation and monitoring of imperfections in the specimens prior to testing.

Based on experiences with rapid prototyping

Within the investigation of form-locking lap joints for paper-tubes, the concept of creating optimal contact by sanding instead of cutting the tubes has been explored. This concept is presented in paragraph 6.4.2.1.

To ease the manufacturing process for intermediate joints, CNC milling technology could be used for wide prefabrication of plug-in joints and pressure forming for sleeve joints.

Structural testing methods

Optimization of the set-up for the 4-point bending test, at the areas of the supports, is possible, with the aim to improve the contact area between the supports and the specimen. Following similar testing methods for round timber profiles, using a curved surface that approximates the size of the sections tested could help.

Monitoring the behaviour of the joints at different areas, for example to analyse the movement of mechanical fixations is also an interesting point for development.

Investigation of testing methods for multi-axial nodes, to approximate realistic scenarios, is a subject that was considered in this research but requires further efforts to build a satisfactory testing set-up.

Structural analysis of paper-tubes

Performing structural tests for tubes with different sections, higher wall thickness and / or diameter comparing to the ones used in this research, is an approach to optimize the results, as in most structural tests the element that failed is the tube.

For the computer simulations on structural analysis, simulating the structural behaviour of a paper tube would be helpful, to approximate the real behaviour of assemblies.

5.2.2 Research, teaching and tutoring

Within the research process a variety of methods were used to study the behaviour of joining techniques for paper-based assemblies and tubular structures. Basic joining principles were approached as a subject for brainstorming within the teaching process (6.4.2.1). In time, there was the opportunity to develop full scale demonstrators (6.3, 6.4.2.2 and 6.4.2.3) and experience the complications of attempting to transfer the current knowledge for building with paper-based products that are not established construction materials. In conclusion, this process was fruitful as it generated active thinking and in combination with hands on approach often delivered answers.

5.2.3 Epilogue

Based on the experiences gained through the whole research process and following the global aim for sustainability, certain considerations are made in the vision of future research on this subject. Following the findings and setting as a priority the aspect of recyclability, a beneficial approach would be to develop further reversible and reusable joints such as case studies 'fiber-reinforced pulp-composite connector', 'timber block node', 'tolerance adaptive timber plug'. In a different way, weighing the durability and structural behaviour of paper-based products, a different approach appears that would match better the concept of building temporarily, also considering the high compatibility with adhesion as a joining method and focusing on the high recyclability of paper products as an asset. This would be to make efforts in researching bio-based binders, to improve the prospects for recycling, which would be applicable for joints similar to the case study 'textile-reinforced epoxy resin laminated joint'.

Literature

Leijten, A. EN1995-1-1: Section 8 - Connections. Retrieved from
https://eurocodes.jrc.ec.europa.eu/doc/WS2008/EN1995_5_Leijten.pdf

McQuaid, M. (2003). *Shigeru Ban*. London N1 9PA: Phaidon Press Limited, Regent's Wharf All Saint's Street.

Miyake, R. (2009). *Shigeru Ban Paper in Architecture*. US: Rizzoli International Publications.

VDI, G. E. K. V. (2004). *Methodical selection of solid connections. Systematic, design catalogues, assistances for work.* Retrieved from

6 Appendix

6.1 'House 1' - Connections

Figure 6.1 Four types of joints are used for the assembly of the skeletal structure.

6.2 Preliminary Experimental Bending Tests

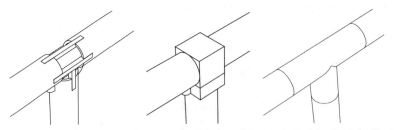

Figure 6.2 Preliminary designs tested a. 'Timber plate puzzle node' - (CW), b. 'Timber block node' - (MW), c. 'Textile-reinforced epoxy resin laminated joint' - (LAMT).

© The Author(s), under exclusive license to Springer Fachmedien Wiesbaden GmbH, part of Springer Nature 2022
E. Kanli, *Experimental Investigations on Joining Techniques for Paper Structures*,
Mechanik, Werkstoffe und Konstruktion im Bauwesen 60,
https://doi.org/10.1007/978-3-658-34501-3_6

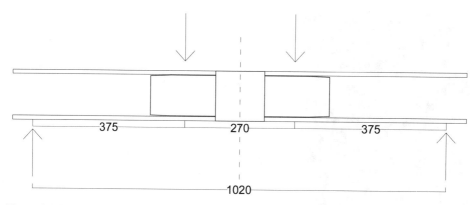

Figure 6.3 Set-up for experimental 4-point bending test

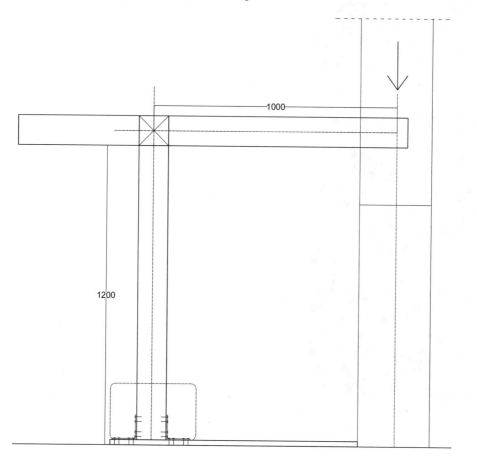

Figure 6.4 Set-up for experimental 90° bending test

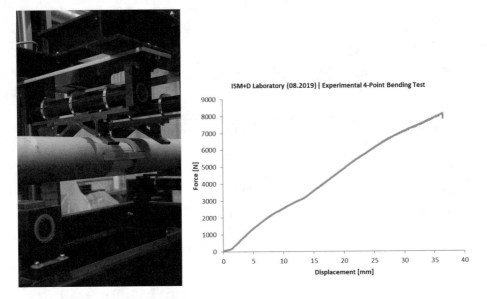

Figure 6.5 'Timber plate puzzle node' (CW) specimen and graph.

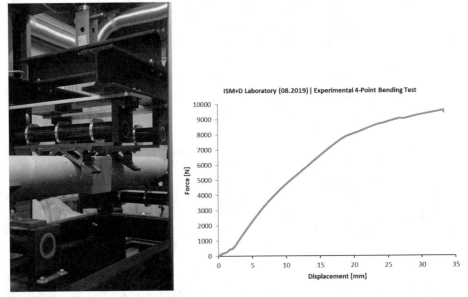

Figure 6.6 'Timber block node' (MW) specimen and graph.

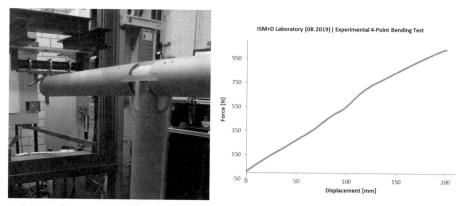

Figure 6.7 'Timber plate puzzle node' (CW) specimen and graph.

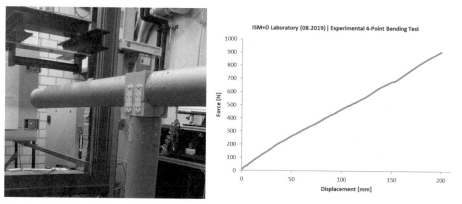

Figure 6.8 'Timber block node' (MW) specimen and graph.

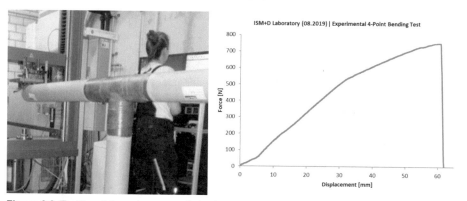

Figure 6.9 'Textile-reinforced epoxy resin laminated joint ' (LAMT) specimen and graph.

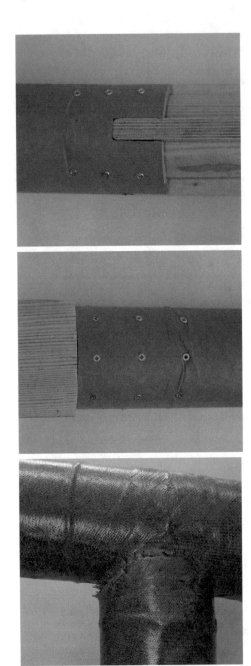

Figure 6.10 Samples of failed specimens

6.3 'Branch-out' Pavilion

Location: Messe Düsseldorf

Time: 10.2018

Contributors:

A group of colleagues from ISM+D contributed to the development and realization of this pavilion with their special skills and expertise, a process that overall required multiple months of work.

The design and execution were planned by MSc students Eva Bodelle and Amir Chhadeh.

The construction details for the timber joints and fundament were worked out in the context of the Master course Façade Technology 2 (see 6.4.2.2).

The process of development is reported in the respective thesis (6.4.1.2).

Figure 6.11 Demonstrator 'branch-out' pavilion.

6.4 Research within education

6.4.1 Supervised Theses and Projects

6.4.1.1 Investigation of joining techniques for paper tubes used as beams in framework

Bachelor thesis

Student: Lyann Perera

Role: Supervisor, Timeframe: 25/09/2018 - 12/02/2019

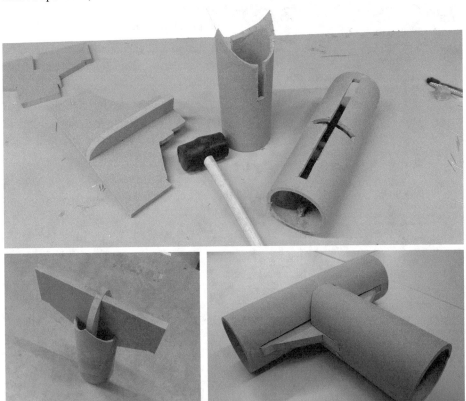

Figure 6.12 Prototype of case study

6.4.1.2 Development of a paper construction for 'glasstec' 2018: design, dimensioning and construction.

Master thesis

Student: Eva-Marie Sarah Jasmin Bodelle

Role: Co-supervision with MSc Nihat Kiziltoprak

Timeframe: 03/01/2019 - 17/07/2019

This thesis reports the process of development for the 'Branch-out' pavilion (see 6.3). The process can be divided mainly in three phases: phase 1 - construction detailing and planning of the installation based on the requirements of the specific location, phase 2 - manufacturing of components, phase 3- build up on-site.

To a different extent the main connecting principle is used as the central aspect studied within the thesis, a wooden joint composed of plates, with reference to the first case study presented in this research (see 6.1). The focus is on the structural performance, observed through FEM analysis and structural testing of full-scale specimens, with the aim to predict its structural behavior effectively.

Figure 6.13 Experimental 4-point bending test.

6.4.1.3 Natural fiber composite joining techniques for paper tubes

International Research exchange program (IREP)

Student: Noah Gilsdorf

Role: Supervisor, Timeframe: 01/05/2019 - 05/08/2019

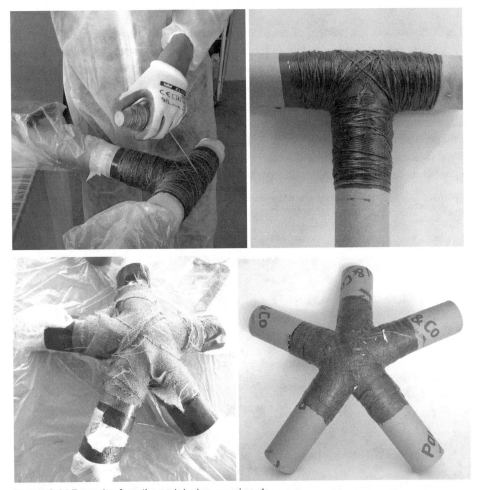

Figure 6.14 Examples from the prototyping experiments.

6.4.2 Educational modules

6.4.2.1 Experimental Façade Technology 1

Bachelor course - Interdisciplinary (Architecture & Civil Engineering)

Role: Coordination, supervision and workshops, Timeframe: 10/2017 - 02/2018

Topic: Development of building systems with use of paper-based products.

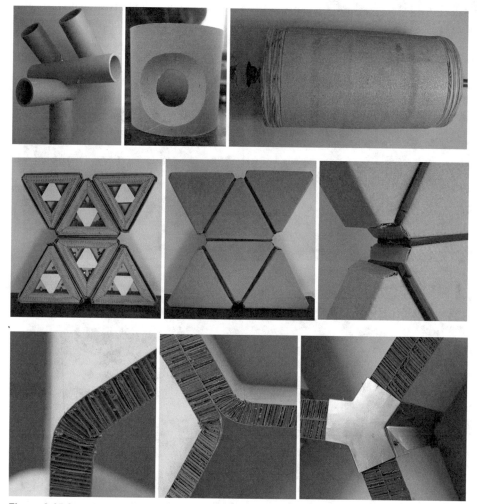

Figure 6.15 Samples of prototypes. Detailing and prototyping are in the main focus.

6.4.2.2 Experimental Façade Technology 2

Master course - Interdisciplinary (Architecture & Civil Engineering)

Role: Coordination, supervision and workshops, Timeframe: 04/2018 - 07/2018

Topic: Elaboration of construction details for the 'branch-out' pavilion

The render on the left presents the design outlines for the pavilion, at the time. Studies on the performance and optimization of the connection points and investigation of solutions for the fundament were the main focus.

Figure 6.16 Frames from the experimental prototypes and development process.

6.4.2.3 Interdisciplinary Project for Civil and Environmental engineers (IPBU)

Master course - Interdisciplinary (Civil Engineering, ISM+D & IFSW[1])

Role: Joint coordination, supervision and workshops, Timeframe: 10/2018 - 02/2019

Topic: 'Phase 1' - Bridge concepts, 'Phase 2' - Realization of selected concept.

Figure 6.17 Frames from the process of development.

---

[1] Prof. Dr.-Ing. Jörg Lange, Dipl.Ing. Aaron van der Heyden

Figure 6.18 Experimental bending test and failure caused by delamination.

6.5 List of figures, tables and graphs

6.5.1 List of figures

1 Research methodology

Figure 1.1 Microscopic view of paperboard (ISM+D)
Figure 1.2 Paper-based products, boards and profiles, that present interest for design applications.
Figure 1.3 The four main clusters of joining techniques: material-closure, force-closure, form-locking and hybrid solutions.
Figure 1.4 Concept for the review of existing solutions.

2 Joining techniques - State of the art

Figure 2.1 A collection of paper-based products fabricated in Germany.
Figure 2.2 A composition of paper-based products that are widely available by the paper-industry.
Figure 2.3 Manufacturing processes.
Figure 2.4 Conceptual construction detail (vertical section, plate-structure).
Figure 2.5 Conceptual construction detail (vertical section: skeletal structure).
Figure 2.6 Conceptual construction detail with focus on the joints, based on fig. 2.4.
Figure 2.7 Conceptual construction detail with focus on the joints based on fig. 2.5.
Figure 2.8 Prototype of a foldable origami-design made by a group of students, 2014, Buck Lab studio (Fall 2014, TU Delft, direct supervisor: Jerzy Latka, head of studio: Marcel Bilow).
Figure 2.9 Series of tensile tests with specimens of laminated paper-board, performed for this research, at the laboratory of ISM+D (02.2019).
Figure 2.10 Examples of common joining techniques for general use.
Figure 2.11 (Left) Packaging product, (right) laser-cut MDF surface.
Figure 2.12 A collection of paper-based boards.
Figure 2.13 Folded edges.
Figure 2.14 (left) 3-dimensional interlocks and (right) plastic screws with wide spiral (right) are both common practices for the assembly of light boards.
Figure 2.15 Detail drawn based on the project: 'Nemunoki Children's Museum' (JPN), 1999, S. Ban.
Figure 2.16 Details drawn based on the project: Apeldoorn Temporary theater (NL), 1992, ABT consulting engineers. The lamella structure is mostly built with paperboard.
Figure 2.17 Project: Apeldoorn Temporary theater (NL), 1992, ABT consulting engineers.
Figure 2.18 Project: Apeldoorn Temporary theater (NL), 1992, ABT consulting engineers.
Figure 2.19 Project: Apeldoorn Temporary theater (NL), 1992, ABT consulting engineers.
Figure 2.20 Project: Apeldoorn Temporary theater (NL), 1992, ABT consulting engineers.
Figure 2.21 The product 'Wikkelhouse', made by the 'Fiction Factory' (Amsterdam (NL).
Figure 2.22 The product 'Wikkelhouse', made by the 'Fiction Factory' (Amsterdam, NL).
Figure 2.23 The product 'Wikkelhouse', made by the 'Fiction Factory' (Amsterdam, NL).
Figure 2.24 A collection of L, U and round paper-based profiles that are readily available in the market.
Figure 2.25 Detail of form-locking joint fixed with tie-wraps, drawn based on project 'Paper Log House', designer S. Ban (2016, in PH).
Figure 2.26 Basic joining techniques briefly investigated in this research: tied rope joint – bamboo style node- (left) and form-fitting pressure joint fixed with bolts (right).

3 Global structural analysis

4 Case studies

5 Global observations

6 Appendix

6.5.2 List of graphs

1 Research methodology (-)

2 Joining techniques - State of the art
Graph 2.1 Experimental tensile test series performed at the laboratory of ISM+D (02.2019).
Graph 2.2 Experimental tensile test series performed at the laboratory of ISM+D (02.2019).

3 Global structural analysis (-)

4 Case studies
Graphs 4.1 Specimen: paper tube (inner diameter 100mm, wall thickness 10mm)
Graph 4.2 Mean curve, extracted from the 3 repetitions (see graph 4.1) and selection of F_{max} Linear.
Graph 4.3 'Configuration 1', testing results, force versus displacement curves.
Graph 4.4 'Configuration 1', average graph line (red color) and linear trend line (black dashed-line).
Graph 4.5 'Configuration 2', testing results, force versus displacement curves.
Graph 4.6 'Configuration 2', average curve.
Graph 4.7 'Configuration 3', testing results, force versus displacement curves.
Graph 4.8 'Configuration 3', average graph line.
Graph 4.9 Pressure-test, force versus displacement curves.
Graph 4.10 Collective data for the case study 'Timber plate puzzle node'. Comparing the mean curves for all three configurations tested and also paper tubes.
Graph 4.11 Testing results for the 'Timber block node', configuration 1.
Graph 4.12 'Timber block node', configuration 1, force versus displacement average curve.
Graph 4.13 Testing results for the 'Timber block node', configuration 2.
Graph 4.14 'Timber block node', configuration 2, force versus displacement average curve.
Graph 4.15 Pressure test, force versus displacement curve.
Graph 4.16 Collective data for case study 'Timber block puzzle node'. Results from 4-point bending tests on configurations 1 and 2 and also paper tubes.
4.2.1.3 Tolerance adaptive joint
Graph 4.17 'Tolerance adaptive timber plug', testing results, force versus displacement curve.
4.2.2.1 Fiber reinforced pulp-composite connector
Graphs 4.18 and 4.19 Vertical displacement and rotational movement versus force.
Graph 4.20 Tensile test, double lap bolted joint (fiber reinforced pulp-composites plates).
4.2.2.2 Textile-reinforced epoxy resin laminated joint
Graph 4.21 'Textile-reinforced epoxy resin laminated joint', configuration 1, force versus displacement curves.
Graph 4.22 'Textile-reinforced epoxy resin laminated joint', configuration 1, force versus displacement average curve.
Graph 4.23 'Textile-reinforced epoxy resin laminated joint', configuration 2, force versus displacement curves.
Graph 4.24 'Textile-reinforced epoxy resin laminated joint', configuration 2, force versus displacement average curve.
4.3 Evaluation of case studies
Graph 4.25 Testing results, force versus displacement average curves.

5 Global observations (-)

6 Appendix (-)

6.5.3 List of tables

5 Global observations

6 Appendix (-)

Printed in the United States
by Baker & Taylor Publisher Services